ANALYSIS WITH
ION-SELECTIVE
ELECTRODES

HEYDEN INTERNATIONAL TOPICS IN
SCIENCE

Editor: L. C. Thomas

ANALYSIS WITH ION-SELECTIVE ELECTRODES

PETER L. BAILEY

Electronic Instruments Limited
Richmond, Surrey

LONDON · NEW YORK · RHEINE

Heyden & Son Ltd., Spectrum House, Alderton Crescent, London NW4 3XX.
Heyden & Son Inc., 225 Park Avenue, New York, N.Y. 10017, U.S.A.
Heyden & Son GmbH, Münsterstrasse 22, 4440 Rheine/Westf., Germany.

ISBN 0 85501 223 4

Printed in Great Britain by Galliard (Printers) Ltd, Gt. Yarmouth, Norfolk.

CONTENTS

FOREWORD

The object of this series of monographs is the timely dissemination of essential information about topics of current interest in science. Interdisciplinary aspects are given fullest attention. The series aims at the presentation of new techniques, ideas and applications in sufficient detail to enable those who are not specialists in a particular subject to appreciate the applicability of the subject matter to their own work, and the bibliographies included in each monograph will guide readers in extending their knowledge of the subject to any desired depth. The depth of treatment of course makes them compact definitive books for the specialist as well. The series will from time to time include more general reviews of selected areas of scientific advancement, for which a somewhat wider readership is envisaged.

The series is to be priced at a level to attract the individual purchaser as much as the librarian. The topics and the depth of treatment should suit both the student and the research worker, academic or industrial. The range of topics in this series will eventually span the whole extent of scientific interests and the authorship will reflect the international nature of the subject matter.

The rapid development of ion-selective electrodes and gas-sensing probes over the past decade, together with the ever increasing range of their applications, reflect the extent to which these devices meet the need for accurate, cheap and rapid analytical and control techniques. Despite this explosion of interest there are still many potential applications to be explored. *Analysis with Ion-Selective Electrodes* fulfils the objective of this Series by making workers in other fields aware of these potentialities. The book covers the theory in sufficient detail to enable readers to assess the applicability of an electrode or probe to their particular analytical problems and to be aware of the most important parameters affecting the electrode's performance. The book should be of value to students, to analysts and to all experimentalists whose work entails analytical procedures. The last chapter, which describes do-it-yourself methods of making electrodes, should be especially valuable and save the cost of the book many times over.

L. C. Thomas

PREFACE

Ion-selective electrodes and gas-sensing probes are still comparatively new analytical tools and there are still many fields of analysis in which their advantages have yet to be exploited. It is the aim of this book to provide the analyst with the necessary information to assess whether an electrode or probe is suitable for use in a particular analysis and to enable him to develop an analytical method which realizes the required accuracy and precision for the analysis. No attempt has been made to present analytical recipes since these inevitably have limited application and an enormous number would have to be given to approach coverage of the field. Furthermore, no full listing of commercial apparatus and its performance is attempted since this information is readily available, free on request, from the relevant manufacturers and is, in any case, all too transitory: many of the sensors produced by different manufacturers for a particular species are based on the same active material and have very similar performances. A certain amount of the theory concerning the mechanism of the operation of the sensors has been included with the intention of giving the reader some insight into the most important parameters affecting their performance. The theoretical section on the ion-selective electrodes based on inorganic salts is more extensive than that on the other types of sensor primarily because, in this case, the theory has been most successful in accounting for the performance.

I am very grateful to Dr M. Riley, who read and commented on the typescript and has helped to iron out many inaccuracies and imprecisions. I also acknowledge gratefully the continuous help of my wife, Kathryn, in the preparation of the manuscript and the diagrams. My thanks are also due to Mr A. E. Bottom for his constructive comments on Chapter 4 and to Mrs J. Edwards for expertly converting my convoluted manuscript into typescript. Not least, I am indebted to Electronic Instruments Limited, without whose help and encouragement the production of this book would not have been possible.

I owe a special debt of gratitude to Prof. E. Pungor and his colleagues at the Technical University in Budapest, who made me very welcome in 1972, aroused my interest in ion-selective electrodes and taught me a great deal about them.

Richmond
July 1976

P. L. Bailey

ACKNOWLEDGEMENTS

Permission is gratefully acknowledged from the Chemical Society, the American Chemical Society, Electronic Instruments Limited, Elsevier Scientific Publishing Company, the Journal of Chemical Education and Pergamon Press for the use of illustrations in this work as follows.

Figures 4.1, 5.2, 5.3, 5.5, 5.6, 5.8, 7.3, 8.2, 8.3, 9.5. 9.6, from *Analytical Chemistry* published by the American Chemical Society.

Figures 6.2, 7.1(c), 10.2, from *Analytica Chimica Acta* published by Elsevier Scientific Publishing Co.

Figures 6.3, 7.2, 9.7, 9.8, from *The Analyst* published by The Chemical Society.

Figure 5.7, from *Talanta* published by Pergamon Press.

Figure 10.3, from *Journal of Chemical Education*.

Figure 5.9, from *Journal of the American Chemical Society*.

Figures 9.1, 9.9(a & c), 9.10 by permission of Electronic Instruments Limited.

CHAPTER 1

INTRODUCTION

Ion-selective electrodes and gas-sensing probes have been the subject of rapidly increasing interest over the past ten years and their development has opened up a large new field of potentiometry. The speed at which this field has developed is a measure of the degree to which the electrodes and probes meet the requirements of the average analyst for rapid, accurate and low-cost analysis. In many frequently performed analyses, such as the determination of nitrate in drinking water, the electrodes and probes are replacing existing techniques; in other cases, such as the on-line monitoring of fluoride in drinking water or sodium ions in boiler feed-waters, the use of a potentiometric sensor has made feasible measurements which are otherwise almost impossible. The growth of these applications has been fast in relation to the growth of other techniques partly because the sensors themselves became widely available commercially soon after their development, and are, in some cases, also quite easy to make in the laboratory. Another contributory factor is that the associated apparatus, such as pH meters and magnetic stirrers, is already part of the standard equipment of most analytical laboratories. Thus many workers have been able to try out the sensors for themselves and many papers on their applications have been published.

The construction and chemical design of the sensors have been steadily improved in the last few years, partly guided by the results of fundamental work on the theory of their operation. In addition to taking much of the trial and error element out of sensor design, these theoretical studies have also helped in the derivation of analytical methods by indicating the optimum conditions of usage of the sensors. Despite these developments, it seems probable that the range of ions which may be directly sensed by ion-selective electrodes of types so far known will not be greatly enlarged. There is, however, a constant trickle of publications describing electrodes which sense more or less exotic determinands, often with rather poor selectivity, and which are extremely useful in specialized applications but unlikely ever to command a sufficiently large market to warrant commercial production. The major recent advance has been the development of new gas-sensing probes which sense dissolved gases in solution.

The performance characteristics of ion-selective electrodes and gas-sensing probes are to be discussed in detail in subsequent chapters, but as an introduction it is worth while summarizing a few of the most important of their features.

Ion-selective electrodes sense the activity, rather than the concentration, of ions in solution; the gas-sensing probes sense the partial pressure of gases in solution. Although the analyst is usually accustomed to measuring the concentration of the species which is to be determined (the determinand) and this is still possible by use of an electrode or probe, a measurement of activity is often also valuable or even preferable; for example, in the analysis of calcium in blood serum, it is the calcium ion activity that is the physiologically important parameter and that must be measured, and not the calcium ion concentration. If the relationship between the activity and concentration of a determinand can be fixed by adding a constant concentration of an inert electrolyte to all samples to swamp out minor variations in sample composition, then an electrode may be used directly for concentration measurements. Details of the various measurement techniques are given in Chapter 9.

Both ion-selective electrodes and gas-sensing probes should ideally respond to determinands according to the Nernst equation. For a determinand X, the Nernst equation for the response of a cell containing an X-selective electrode may be written

$$E = E^\circ \pm \frac{RT}{zF} \ln a_X \qquad (1.1)$$

where E° is the standard potential of the cell, a_X is the activity of ion X and z is the number of charges on X. The sign in the equation is positive when X is a cation and negative when it is an anion. An analogous equation for the response of gas-sensing probes is given in Chapter 7. Equation (1.1) describes the pure response of the electrode to X and thus assumes that there is no other ion in the solution to which the electrode can respond. Unfortunately this ideal selectivity is not a feature of real electrodes, as is discussed in Chapter 3. Furthermore, the range over which an electrode will give a Nernstian response to the determinand X is limited in practice, even in the absence of other, interfering, ions to which the electrode responds (interferents). It may be noted also that it has become common to refer to the response of an electrode or probe as Nernstian whenever the electrode potential is proportional to the logarithm of the determinand ion activity, $\ln a_X$, even if the constant of proportionality is not exactly RT/zF. For $z = 1$, RT/zF is equal to 59 mV at 25 °C, but values down to even 55 mV are frequently referred to as Nernstian, although strictly this is incorrect. The term 'sub-Nernstian' is used to describe electrode responses when the potential is proportional to ln a_X, but the proportionality constant is even lower than about 55 mV. Such low responses usually indicate a faulty electrode and irreproducible readings may be anticipated.

Since the electrodes and probes give a voltage signal which is proportional to the logarithm of the determinand activity (Eq. 1.1), the accuracy and precision of the determinations are sometimes poorer than for other techniques, especially when high determinand activities are measured. However, the accuracy and precision of the determinations may remain essentially unaltered over several decades of activity. The accuracy and precision attainable will, of course, depend on the degree of control exercised over the experimental conditions, such as temperature and stirring rate, and on the measurement technique. In direct determinations the ion-selective electrode and a reference electrode are immersed in a stirred sample and the resultant cell potential related to the determinand activity or concentration by means of a calibration graph; the expected accuracy and precision for such a straightforward method should be comparable to those of a pH measurement. The 'glass electrode' or hydrogen ion-selective electrode is the best-behaved of all ion-selective electrodes, yet measurements of pH to better than ± 0.01 pH are generally accepted to require considerable care; but, ± 0.01 pH is equivalent to ± 0.6 mV or, in terms of hydrogen ion activity, $\pm 2\%$. Thus, with care an uncertainty of $\pm 2\%$ or even down to $\pm 0.5\%$ may be obtained; this may be good in comparison with other techniques at the 10^{-5} M level but is often poor at 10^{-2} M. The accuracy and precision may be improved by resorting to a more elaborate technique involving several potential readings, such as the standard addition technique (Chapter 9); these techniques reduce the effect on the result of errors in single readings. Consequently a balance must be struck between accuracy and convenience in the choice of technique. A corollary of this is that the sensors are more suited to the relatively imprecise analysis of samples containing the determinand in a wide range of activity (or concentration) than to precise analysis of samples containing the determinand in a narrow range.

One of the most attractive features of the electrodes and probes is the speed with which they permit a sample to be analysed, and the ease with which the methods may be made semi-automatic or fully automatic. If samples are measured by a direct method, all that has to be done is to pretreat the sample by addition of an aliquot of a pretreatment reagent (Chapter 3), stir the sample and put the gas-sensing probe or ion-selective electrode and reference electrode into it; the equilibrium cell potential may be read usually within about a minute and the answer obtained. Thus about 20–30 samples per hour may be analysed manually; if the sensor is incorporated in a flow system, discrete samples may be analysed at rates up to about 60 per hour or a sample stream may be analysed continuously.

Other advantages of analyses by electrodes or probes are that the methods may be non-destructive and are adaptable to very small sample volumes (e.g. fluoride samples have been analysed[1] with volumes down to 10 μl). Furthermore, analyses may be made without difficulty of highly coloured, viscous samples containing a high concentration of suspended solids. However, problems do arise in the analysis of some non-aqueous and partially non-aqueous samples.

TERMINOLOGY

Before further discussion of the performance of ion-selective electrodes and gas-sensing probes, some of the terms used must be defined or explained.

Several of the terms associated with ion-selective electrodes and gas-sensing probes have accumulated a surprising number of variants during their short history. In most cases one version is clearly preferable as it describes the parameter or object concerned more precisely than the other variants; in other cases the choice may be more arbitrary. A recent IUPAC bulletin of recommendations for nomenclature[2] is helpful in sorting out some of the terms, but bears the unmistakable hallmark of a document produced as a compromise by a committee. Some of the more general terms to be used in this book are explained in the following sections.

Ion-selective electrodes. These may be simply defined as electrochemical sensors which respond to ionic activities according to Eq. (1.1), sometimes with sub-theoretical slopes. The electrodes do not directly sense redox potentials as do conventional electrodes of the first and second kind (e.g. the silver metal electrode which has been used to sense silver ions or the silver/silver chloride electrode which senses chloride ions). The terms 'specific ion electrode' or 'ion-specific electrode' are incorrect since they imply that the electrode responds only to one ion, which is never true. Further alternative terms include 'selective ion-sensitive electrode', which is linguistically correct but unnecessarily unwieldy, and 'ion-selective membrane electrode', which, since all ion-selective electrodes have membranes, gives no extra information.

Gas-sensing probes. These probes are complete electrochemical cells, incorporating their own sensing electrode and reference electrode. They respond to the partial pressure of gases in solution according to the Nernst equation. The probes are in two categories. First there are gas-sensing membrane probes which have a gas-permeable hydrophobic membrane on the end to separate the sample from the internal sensing electrode; these probes are immersed in the sample during measurements. The second category is of gas-sensing probes without membrane, which are suspended above samples during measurements; the sensing electrode of the probe is thus separated from the sample by a gap of air (giving rise to the nickname 'air-gap electrode'). Both types of gas-sensing probe are illustrated in Fig. 7.1.

The terms 'gas electrode' and 'gas-sensing electrode' (the provisional IUPAC recommendation) have also been used to describe the probes, but are unsatisfactory as the probes are complete electrochemical cells rather than electrodes in any accepted meaning of the word (no mean distinction). The term 'probe', although not giving any information concerning the operation of the devices, does imply a complete system, and is brief and not incorrect.

Primary ion. The primary ion is the ion that the electrode is designed to measure; for example, the sodium ion is the primary ion sensed by a sodium electrode. This does not necessarily imply that the electrode is most responsive to the primary ion, since nearly all sodium electrodes respond to hydrogen and silver ions more strongly than to sodium ions.

Determinand. The species to be determined in an analysis is termed the determinand.[3] In direct potentiometric methods (Chapter 9) the determinand is nearly always the primary ion, but this is not necessarily so in indirect methods such as those involving potentiometric titrations.

Response time. The response time of an ion-selective electrode or gas-sensing probe is defined by IUPAC as the time taken for the potential of the cell containing the electrode to reach a value 1 mV from the final equilibrium potential after a supposedly instantaneous change in determinand activity (or partial pressure for a probe). The response time is extremely dependent on the experimental conditions; for a quoted value to be useful, therefore, it is essential to know the initial and final determinand activities, stirring rate and the way in which the activity was changed.

Interferent. An interferent is defined as any species, other than the primary ion to which the electrode or probe responds. The presence of a significant concentration of interferent in a sample leads to the apparent concentration of primary ion being erroneously high.

Selectivity coefficient. The selectivity coefficient k_{AB}, which is defined by Eq. (3.10), is a measure of the selectivity of an electrode for the primary ion A (usually the determinand) in the presence of the interferent B. The smaller the value of k_{AB} the more selective the electrode for A in the presence of B.

The selectivity coefficient has also been called the selectivity constant and the selectivity factor. The former term is inappropriate because selectivity coefficients are seldom even approximately constant for all experimental conditions and ratios of A to B. The latter term, while being as accurate as 'selectivity coefficient', has not received wide acceptance.

Limit of detection. The meaning of the term 'limit of detection' is not sufficiently straightforward for it to be summarized in a few words, and therefore the explanation is left to be dealt with in Chapter 3.

Calibration graph. A calibration graph for an ion-selective electrode is accepted as meaning a plot of the potential difference between the ion-selective electrode and a reference electrode against the logarithm of the activity or concentration of the primary ion in the measurement cell. For a primary ion A, the logarithm of its activity or concentration is usually plotted along the abscissa of the graph in units of pA ($= -\log_{10} a_A$ or $-\log_{10}$ [A] as appropriate) and the cell potential is plotted along the ordinate. In the case of a gas-sensing probe, the probe potential is plotted against the logarithm of the concentration of the sensed species in either the ionic or dissolved gaseous form (e.g. against pNH_4 or pNH_3).

CLASSIFICATION OF SENSORS

Many different classifications of ion-selective electrodes have been presented, of which the most logical, from both practical and theoretical considerations, is that based on the type of active material used to make the membrane. If this system is used, the majority of ion-selective electrodes fall easily into three main classes, and a fourth may be used to account for the rest. It is on this basis that the electrodes have been divided between Chapters 4, 5, 6 and 8, while the gas-sensing probes (Chapter 7), which are quite distinct, fall into a class of their own. Thus five classes of sensor may be identified as follows.

1. *Glass electrodes.* Electrodes with glass membranes were well known long before the recent development of the other classes of ion-selective electrode. The hydrogen ion-selective electrode is the most thoroughly investigated and highly developed of all ion-selective electrodes; because of the extreme properties of hydrogen ions, in particular their very high mobility, this electrode is extraordinarily well-behaved, having very high selectivity ($k_{HNa} \simeq 10^{-14}$) and a long response range (pH 0 to 14). Subsequently, glass electrodes for the measurement of Na^+, K^+, $NH_4{}^+$ and some other cations have been developed and are discussed in Chapter 4. Electrodes made from chalcogenide glasses sensitive to Fe^{3+} and Cu^{2+} are mentioned in Chapter 8.

2. *Electrodes based on inorganic salts.* Into this class fall the electrodes based on the silver halides, silver sulfide, lanthanum fluoride and heavy metal sulfides. The membranes of these electrodes have been produced in a variety of homogeneous and heterogeneous forms ranging from single crystals to dispersions of the active material in an inert matrix such as silicone rubber or polythene. Frequently there is little difference in performance between the various forms of membrane based on a particular active material. Ions measured by electrodes in this class include F^-, Cl^-, Br^-, I^-, CN^-, S^{2-}, Ag^+, Cu^{2+} and Pb^{2+}: these electrodes are described in Chapter 5.

3. *Electrodes based on organic ion exchangers and neutral carriers.* Organic ion exchangers are used to make electrodes with liquid or solid membranes which are selective to ions such as Ca^{2+}, $NO_3{}^-$ and $ClO_4{}^-$. The cation-selective electrodes use cation exchangers or complexing agents as active material, whereas the anion-selective electrodes use anion exchangers. Electrodes based on neutral carriers generally have higher selectivity and are better-behaved than the other electrodes in this class: the most widely applied of these electrodes are

those selective to K^+ and NH_4^+. Both types of electrode in this class are described in Chapter 6.

4. *Gas-sensing probes.* The gas-sensing probes are complete electrochemical cells and thus quite distinct in their principles of operation and performance from ion-selective electrodes. The gas-sensing membrane probe which senses CO_2 has been in increasingly common use in medical applications since 1957. Recently the range of determinands which may be sensed by these probes has been extended to include dissolved gases such as NH_3 and SO_2. Gas-sensing probes are described in Chapter 7.

5. *Miscellaneous electrodes.* A number of types of sensor have been described which do not come within the other classes, despite similarities in some cases; these types are grouped together for convenience and are briefly described in Chapter 8. Sensors in this class include the enzyme electrodes (which are somewhat similar in concept to the gas-sensing probes), the surfactant electrodes (which are similar to the electrodes based on organic ion exchangers) and ion-sensitive field effect transistors (ISFETs).

HISTORY OF THE DEVELOPMENT OF ION-SELECTIVE ELECTRODES AND GAS-SENSING PROBES

To conclude this introductory chapter, it is worth while to recount briefly the most important stages in the history of the development of ion-selective electrodes and gas-sensing probes.

The history starts with the observations of Cremer[4] in 1906 and the more detailed investigations of Haber and Klemensiewicz,[5] which showed that a hydrogen ion-sensing glass electrode responded to hydrogen ions according to the Nernst equation. Hughes[6] later compared the glass electrode with the hydrogen electrode and adumbrated the development of the other cation-selective electrodes in his experiments on the interference of cations on glasses containing different amounts of Al_2O_3. In 1934, Lengyel and Blum[7] showed that an electrode made from a glass containing Al_2O_3 or B_2O_3 could give a Nernstian response to sodium ions. Subsequently, mostly in the late 1950s and early 1960s, the research teams of Nicolsky in the U.S.S.R. and Eisenman in the U.S.A. studied the properties of many families of glasses and also deduced theories to account for the origin of the glass electrode potential. A review of the development of the compositions of glasses sensitive to hydrogen ions and other cations has been presented by Isard[8] in a chapter of the very useful book edited by Eisenman.[9] The theory of glass electrodes is also extensively covered in this book, although for more recent accounts of developments in this rapidly changing field the biennial reviews in *Analytical Chemistry* are useful.

The first attempts to make ion-selective electrodes from materials other than glass were made by Tendeloo[10,11] and by Kolthoff and Sanders[12] in the 1930s. These latter authors prepared halide-selective electrodes from pellets of silver halide. The potential advantages of these electrodes over the redox-sensitive electrodes of the second kind were appreciated at that time, but unfortunately the performance of the electrodes was not sufficiently good. In the late 1950s Tendeloo and Krips[13,14] produced a calcium-sensing electrode from a membrane of paraffin impregnated with calcium oxalate, but the performance of this electrode was not satisfactory in several respects. Pungor and Hollós-Rokosinyi[15] produced an electrode membrane by embedding silver iodide in paraffin, and demonstrated the selectivity of the electrode to iodide ions in the presence of chloride ions. Following this, Pungor and his co-workers developed a whole range of electrodes of good selectivity and giving Nernstian response; the electrodes all had heterogeneous membranes consisting of an active material supported in an inert matrix of silicone rubber. These electrodes were the first truly selective ion-selective electrodes (non-glass) to be developed which gave a response which is thermodynamically reversible with respect to the primary ion. The silver and halide ion-selective electrodes became commercially available in Hungary in 1965.

In the following year another breakthrough was made with the development by Frant and Ross[16] of a highly selective electrode to measure fluoride ions. Because of (1) the importance of fluoride determination, primarily in the analysis of potable waters, (2) the difficulties associated with alternative analytical methods and (3) the very satisfactory performance of the electrode, it received wide acceptance very rapidly. The membrane of this electrode consists of a crystal of lanthanum fluoride doped with europium fluoride. Further electrodes based on inorganic salts were later developed[17] to measure the activities of copper, lead and cadmium ions.

In 1967 Ross[18] announced the first of the electrodes with a liquid membrane; this electrode is used for measuring calcium ion activity and has as active material a phosphate ester. Immediate interest was aroused in the field of biomedical research by the advent of this electrode because it permitted calcium ion activity to be measured directly in blood serum. Subsequently a special version of this electrode was manufactured for measurements in blood serum. Soon after the appearance of the calcium electrode other electrodes with liquid membranes based on organic ion exchangers were produced. The most important of these electrodes is the nitrate electrode, which permits nitrate concentrations in waters to be measured much more simply and rapidly than by existing colorimetric methods. The nitrate electrode is now widely used for this purpose, both for the analysis of discrete samples in the laboratory and also, when incorporated in an automatic analyser, for the continuous monitoring of rivers, water supplies, etc.

The work of Stefanac and Simon[19,20] on the alkali ion selectivity of membranes containing electrically neutral antibiotics led to development of the potassium-selective electrodes based on valinomycin; the performance of the first commercial version of this electrode was reported by Pioda, Stankova and Simon[21] in 1969. This electrode shows far greater selectivity for potassium ions in the presence of sodium ions than the potassium-selective glass electrode. Other electrodes based on electrically neutral carriers have been prepared for, e.g. the determination of ammonium and calcium ions. A feature of the development of these electrodes is the way in which the theory of their operation has been built up in parallel with the experimental work and the theory has been successfully used to predict suitable membrane materials and electrode performance characteristics (see, e.g. Refs 22 and 23).

The first of the gas-sensing probes was the carbon dioxide sensing probe specifically designed for measuring the partial pressure of carbon dioxide in blood. This was first described by Stow, Baer and Randall[24] and later substantially developed by Severinghaus and Bradley.[25] A decade later, the ammonia probe was the next gas-sensing probe to appear. This probe, which has a wide range of applications in the analysis of samples from sewage effluents and river water to Kjeldahl digests, is now well established. More recently, other gas-sensing probes for the measurement of sulfur dioxide, nitrogen oxides and hydrogen sulfide have been developed.

A full list of the most important papers from the historical point of view is given in the review article by Covington;[26] additionally, Pungor and Tóth[27] have reviewed the development of the electrodes based on inorganic salts. Comprehensive coverage of the more recent progress in the applications and design of ion-selective electrodes and gas-sensing probes is given in the biennial reviews in *Analytical Chemistry*.

REFERENCES

1. R. A. Durst, *Anal. Chem.* **40**, 931 (1968).
2. *Recommendations for Nomenclature of Ion-Selective Electrodes*, Appendices on Provisional Nomenclature, Symbols, Units and Standards—Number 43, IUPAC Secretariat, Oxford (January 1975).
3. A. L. Wilson, *Talanta* **12**, 701 (1965).
4. M. Cremer, *Z. Biol.* **47**, 562 (1906).
5. F. Haber and Z. Klemensiewicz, *Physik. Chem. (Leipzig)* **67**, 385 (1909).
6. W. S. Hughes, *J. Am. Chem. Soc.* **44**, 2860 (1922).
7. B. Lengyel and E. Blum, *Trans. Faraday Soc.* **30**, 461 (1934).
8. J. O. Isard, Chapter 3 in Ref. 9.
9. G. Eisenman (ed.), *Glass Electrodes for Hydrogen and other Cations*, Marcel Dekker, New York (1967).
10. H. J. C. Tendeloo, *J. Biol. Chem.* **113**, 333 (1936).
11. H. J. C. Tendeloo, *Rec. Trav. Chim.* **55**, 227 (1936).
12. I. M. Kolthoff and H. L. Sanders, *J. Am. Chem. Soc.* **59**, 416 (1937).
13. H. J. C. Tendeloo and A. Krips, *Rec. Trav. Chim.* **76**, 703 (1957).
14. H. J. C. Tendeloo and A. Krips, *Rec. Trav. Chim.* **76**, 946 (1957).
15. E. Pungor and Hollós-Rokosinyi, *Acta Chim. Acad. Sci. Hung.* **27**, 63 (1961).
16. M. S. Frant and J. W. Ross, *Science, N.Y.* **154**, 3756 (1966).
17. J. W. Ross, in *Ion-Selective Electrodes*, R. A. Durst, editor, Chapter 2. N.B.S. Spec. Publ. No. 314, Washington, D.C. (1969).
18. J. W. Ross, *Science, N.Y.* **156**, 3780 (1967).
19. Z. Stefanac and W. Simon, *Chimia* **20**, 436 (1966).

20. Z. Stefanac and W. Simon, *Microchem. J.* **12,** 125 (1967).
21. L. A. R. Pioda, V. Stankova and W. Simon, *Anal. Lett.* **2,** 665 (1969).
22. W. Simon, W. E. Morf and P. C. Meier, *Structure and Bonding* **16,** 113 (1973).
23. W. E. Morf, E. Lindner and W. Simon, *Anal. Chem.* **47,** 1596 (1975).
24. R. F. Stow, R. F. Baer and B. F. Randall, *Arch. Phys. Med. Rehabil.* **38,** 646 (1957).
25. W. Severinghaus and A. F. Bradley, *J. Appl. Physiol.* **13,** 515 (1958).
26. A. K. Covington, *CRC Crit. Rev. Anal. Chem.* 355 (1974).
27. E. Pungor and K. Tóth, *Pure Appl. Chem.* **34,** 105 (1973).

CHAPTER 2

CELLS AND REFERENCE ELECTRODES

ELECTROCHEMICAL CELLS

To make a measurement with an ion-selective electrode, an electrochemical cell must be set up. The constitution of this cell markedly affects its performance and the analytical measurements made with it, and thus care must be taken in selecting the components. Three types of cell are commonly used in work with ion-selective electrodes

Cell 1. Cells without liquid junctions

Reference electrode ‖ sample ‖ ion-selective electrode

Cell 2. Cells with one liquid junction, J (i.e. with single junction reference electrodes)

J

Reference electrode | bridge solution ‖ sample ‖ ion-selective electrode

Reference electrode assembly

Cell 3. Cells with two liquid junctions, J_1 and J_2 (i.e. with double junction reference electrodes)

J_1 J_2

Reference electrode | reference solution | bridge solution ‖ sample ‖ ion-selective electrode

Reference electrode assembly

Thus the cells consist of three components: the reference electrode assembly, the sample and the ion-selective electrode.

In cell 1 the reference electrode is simply another ion-selective electrode, which is selective for a species different from the indicator electrode. In cells 2 and 3 the reference electrode, liquid junction and bridge and reference solutions are usually contained in a single reference electrode assembly, the whole being commonly known as a reference electrode (a usage which will be adopted in other chapters). The potential stability of most reference electrode assemblies is usually sufficient for simple pH measurements; however, the requirements for reference electrode performance in ion-selective electrode work are more exacting because the cell potentials must be determined more precisely, and sample contamination by bridge solution is more likely to be a problem. If the conditions of measurements involve, for example, thermal cycling or variations in sample acidity, then the correct choice of reference electrode assembly becomes even more important than usual.

The function of the reference electrode is to act as a half-cell of constant potential against which variations in the potential of the ion-selective electrode in various samples can be measured. No reference electrode adopts an absolutely constant potential in all conditions and thus one must be chosen which behaves sufficiently ideally for the required measurement. The potential of the reference electrode assembly as a whole will consist of the reference electrode potential plus the potentials across the liquid junctions. Instability of these liquid junction potentials is an important source of error in ion-selective electrode work and will be considered later in the chapter. Such errors may, of course, be completely eliminated by using cells without liquid junctions (cell 1).

The next component of the cell to be considered is the sample or test portion. This may either be in its original state or have been altered by the addition of preservatives, buffers, ionic strength adjusters, complexing or decomplexing reagents. Sample pretreatment will be discussed in later chapters.

The final component of the cell is the ion-selective electrode itself. The choice of ion-selective electrode for a given measurement may lie between types working on different principles, or between electrodes working on the same principle but having different mechanical designs. For many electrodes there is a further choice of whether to use a commercial electrode or a home-made electrode.

In this chapter the components of the cell other than the ion-selective electrode and sample will be considered in turn, and the choice of cell types for a given measurement will be discussed. Consideration of gas-sensing probes, which are themselves complete electrochemical cells, is left to Chapter 7.

Cells without liquid junctions

For the most accurate work cells without liquid junctions should be used to avoid uncertainties associated with liquid junctions. In such cells almost any satisfactory ion-selective electrode may be used as reference electrode, provided that, in addition to the determinand, there is a constant activity of the ions which the reference electrode senses in the sample. If this constant activity of ions is not already present, it can be added during sample pretreatment. This technique has considerable advantages over conventional techniques in terms of precision of measurement. A practical limitation, however, is that most ion-selective electrodes have a high impedance and most pH meters have only one high-impedance input; thus either a special pH meter with two high-impedance inputs must be used or it must be ensured that one of the electrodes has a low impedance ($<$ about 10^4 Ω). A further limitation is that samples must be free from substances that interfere with either electrode.

Several examples of the use of such cells are reported and they indicate the power of the technique. Manahan[1] added a constant level of fluoride ions to samples of nitrate and used a fluoride electrode as reference electrode in a cell in conjunction with a nitrate electrode. He reported that the resulting system was remarkably stable, with short-term fluctuations of less than ± 0.02 mV, which with care should be improvable to ± 0.01 mV. Mertens and Massart[2] have critically evaluated this method for the determination of nitrate in samples of mineral waters and have made clear the limitations due to pH (which must be kept between 6 and 8) and interferents. When applying the same technique to samples containing allyl alcohol or acetonitrile, Stelting and Manahan[3] found that the solvent effects at the reference electrode produced greater errors than a liquid junction.

The silver/silver chloride electrode has been commonly used as reference electrode in cells with samples containing a constant background activity of chloride ions; in the most-studied system, used to establish the pH scale, the ion-selective electrode was a hydrogen ion-selective electrode, such as the hydrogen electrode or the glass electrode (see the discussion by Bates[4]).

Cells without liquid junctions are used in a new industrial analyser.[5] For the measurement of fluoride, for example, the two electrodes used in the cell of the analyser are the fluoride electrode and a sodium electrode. The sodium electrode responds to the activity of sodium ions in the treated sample, which largely results from the buffer added to the sample. Provided that this activity remains much larger than any variable sodium ion activity in the untreated sample, and that the buffer is strong enough to eliminate pH variations in the treated sample, which might produce variable hydrogen ion interference, the sodium electrode adopts a reference potential against which the fluoride electrode potential may be measured. The analyser uses an amplifier with two high impedance inputs.

Cells with liquid junctions

In cells with liquid junctions (cells 2 and 3), in contrast to cells without liquid junctions, the reference electrode assembly adopts a potential which is assumed to be independent of the composition of the sample. This involves the assumptions that the potentials of the reference electrode and liquid junctions are constant. Such assumptions are nearly always sufficiently good if suitable choices of reference electrode, bridge solution and liquid junction are made. The ranges of these which are commonly available are discussed in subsequent sections.

Analytical methods using cells with liquid junctions are generally easier to operate as a result of this independence of the potential of the reference electrode from variations in sample composition. In contrast, it is often difficult to develop a satisfactory sample pretreatment for a method using a cell without liquid junction which will bring the sample to the optimum condition for the response of the two electrodes and also keep the potential of the reference electrode constant and stable. Other advantages of the cells with liquid junctions are that commercial reference electrode assemblies are much cheaper than ion-selective electrodes, last almost indefinitely in normal laboratory applications, if treated carefully, and require very little maintenance. Consequently cells with liquid junctions are nearly always used in analytical work, while cells without liquid junctions are usually reserved for physicochemical research.

REFERENCE ELECTRODES

Introduction

The essential characteristics of a satisfactory reference electrode are stability, reproducibility and reversibility.[6] These characteristics are all interrelated and, in practice, not perfectly realizable. The magnitude of the departure from ideality which may be tolerated for a reference electrode in a cell for a given measurement must be assessed in relation to the acceptable inaccuracy and imprecision of the determination and to the expected departure from ideality of the ion-selective electrode.

The stability of the reference electrode will be affected by both chemical and physical factors. Ideally, the chemical components of the electrode should be stable within the assembly and should not react with species in the sample. In practice, however, samples often contain species which oxidize, complex or otherwise react with the components of the electrode, and thus would cause potential drift or shift if they reached the electrode. To prevent this a bridge solution, often of strong potassium chloride solution, is interposed between the reference electrode and the sample; the electrode assembly is so designed that the bridge solution continuously flows across the junction between this solution and the sample, thus stopping diffusion of sample species to the electrode. For high stability, the reference electrode should have a low temperature coefficient and be insensitive to variations in incident light, pressure and other physical parameters.

The reproducibility of a reference electrode is closely associated with its thermodynamic reversibility and the magnitude of the exchange current of the electrode reaction. A perfectly reproducible electrode should obey the Nernst equation, show no temperature or concentration hysteresis and regain equilibrium rapidly after polarization (i.e. after passage of current). Polarization, however, is not a problem when the electrode is connected to a modern pH meter, since the current flowing is extremely small.

For routine pH measurements to ± 0.05 pH (± 3 mV) detailed consideration of the performance of reference electrodes is largely irrelevant, and any one of a large range of commercial electrodes may be used. But with ion-selective electrodes, particularly when they are used in direct potentiometry (Chapter 9), measurements are often made with a precision of ± 0.002 pX units (i.e. ± 0.1 mV or $\pm 0.4\%$ a_X) and such considerations cannot then be ignored. However, it is frequently satisfactory to use one or other of the common reference electrodes if adequate precautions are taken and the limitations are appreciated. It should be emphasized that the value of the potential of the reference electrode is not normally a criterion of selection; only the constancy of that potential under the conditions of measurement is important.

There are four common reference electrodes. The properties and applications of each will be dealt with in turn. The preparation and properties of these and other reference electrodes were comprehensively reviewed in 1961 in the book edited by Ives and Janz.[7] More recent work has been reviewed by Bates[8] and by Covington.[9] Mattock[10] has written a useful review of reference electrodes which emphasizes practical points.

The calomel electrode

The calomel reference electrode is the commonest of all reference electrodes. It consists of a pool of mercury covered by a layer of mercurous chloride (calomel) in contact with a reference solution containing chloride ions and saturated with mercurous chloride. The solution chosen is nearly always of potassium chloride, although hydrochloric acid has been used in fundamental studies of the electrode. The electrode gives a Nernstian response to chloride ions. The strength of the potassium chloride solution used gives the electrode its popular name; thus saturated, 3.8 M and 3.5 M potassium chloride solutions in contact with the mercurous chloride layer give rise to electrodes termed saturated calomel electrode (SCE), 3.8 M and 3.5 M calomel electrode, respectively.

The use of potassium chloride solution as the source of chloride ions is appropriate because it gives rise to small liquid junction potentials at the electrode assembly's liquid junction with the sample (see p. 27). It thus fulfils the conditions for being both a suitable reference solution and a bridge solution; furthermore, it is readily and cheaply available in a pure state. Another advantage is that the mercurous chloride has only a very low solubility in it regardless of concentration.

The half-cell reaction is

$$Hg + Cl^- \rightleftharpoons \tfrac{1}{2} Hg_2Cl_2 + e$$

The performance of the cell may be assessed in terms of the criteria already outlined: stability, reproducibility and reversibility.

The components of the cell are chemically stable with the exception of the mercurous chloride, which disproportionates at a significant rate, at temperatures above 70 °C, according to the equation

$$Hg_2Cl_2 \rightleftharpoons Hg + HgCl_2 \tag{2.1}$$

This disproportionation gives the electrode a progressively shorter life as the temperature is raised, and excessive drift soon makes the electrode useless above 100 °C.[11] At the other end of the temperature scale, the electrode may be used at temperatures from 0 °C down to −30 °C if 50% glycerol is added to the bridge solution.[12]

The oxidation of mercury by dissolved oxygen is an important effect in the calomel/hydrochloric acid electrodes; this causes drift unless oxygen is scrupulously excluded. However, in the usual calomel/potassium chloride electrodes this effect is not significant.

As previously indicated, interference from species diffusing from the sample is usually prevented by using a head of bridge solution which maintains a steady flow of the bridge solution into the sample. However, impurities in the bridge solution itself may also affect the electrode. The most common interference is from bromide ions. Pinching and Bates[13] have shown that the calomel electrode is very sensitive to bromide ions; they have described a method for reducing the bromide impurities in potassium chloride to negligible proportions (≤ 0.001 mol %). It is not usually worth while, when one uses the calomel electrode as a reference electrode of unknown but fixed potential, to worry about interferents such as traces of bromide in the bridge solution: they produce a small shift in the electrode potential, rather than drift, and will not seriously affect the electrode's stability, although small changes in potential will occur whenever the bridge solution is replaced unless the trace impurities remain at constant concentration. Other species which would interfere if allowed to reach the electrode include redox agents, sulfide ions and complexants such as ethylenediamine tetra-acetate (EDTA).

The most unsatisfactory feature of the performance of the calomel electrode is its reaction to temperature changes with, in particular, thermal hysteresis effects. Much has been written on this subject and for many people it is the reason they prefer other types of reference electrode; yet most of the real difficulties are seldom met in practice or are not as serious as they initially appear. A detailed discussion of this subject is presented by Mattock.[10]

It has been shown by Mattock[10] that the magnitude of thermal hysteresis effects is strongly influenced by the method of manufacture in the case of saturated calomel electrodes. Electrodes prepared with the mercurous chloride in contact with saturated potassium chloride maintained at saturation by a remote source of solid potassium chloride showed virtually no hysteresis, in line with the behaviour of unsaturated calomel electrodes; however, electrodes prepared with potassium chloride crystals in contact with the mercurous chloride showed considerable hysteresis. It appears that lack of true thermal equilibrium within the electrode and relatively slow reversal of reaction (2.1) are the main problems, and these can both be minimized by good cell design. However, in most ion-selective electrode work thermal hysteresis need not be considered since both the reference electrode and sample are at approximately 20 °C, and virtually never cycled to and from high temperatures. Thus the hysteresis problems of calomel electrodes, although important in some unusual applications, may normally be neglected. Use of an unsaturated electrode with a mechanical design ensuring fast approach to thermal equilibrium minimizes the effect when it is significant in a given application; but it is usually best in such cases to use a silver/silver chloride or 'Thalamid' electrode instead of the calomel electrode. Alternatively, a remote calomel electrode held at constant temperature may be used and joined to the sample solution by a salt bridge packed with cotton wool or similar material to prevent convection.[14]

The temperature coefficient of a calomel electrode may be expressed by either the isothermal or the non-isothermal value. The isothermal temperature coefficient of the calomel electrode is measured in the following cell (cell 4):

Cell 4

Reference electrode assembly

$$\text{Pt, H}_2(g) \mid \text{solution} \underset{T_1}{\parallel} \qquad \underset{T_1}{\text{KCl} \mid \text{Hg}_2\text{Cl}_2, \text{Hg}}$$

whereas the non-isothermal temperature coefficient is measured in the following cell (cell 5):

Cell 5

Reference electrode assembly   Reference electrode assembly

$$\underset{T_1}{\text{Hg, Hg}_2\text{Cl}_2 \mid \text{KCl}} \quad \parallel \text{KCl} \parallel \quad \underset{T_2}{\text{KCl} \mid \text{Hg}_2\text{Cl}_2, \text{Hg}}$$

Cell 5 contains a bridge of potassium chloride solution, of the same concentration as in the two reference electrode assemblies, with a temperature gradient from T_1 to T_2 between the ends.

The isothermal temperature coefficient is thus the difference in the rate of change of potential of the calomel and hydrogen electrodes with change of temperature; although the potential of the hydrogen electrode is defined as zero at all temperatures, it has a non-isothermal temperature coefficient of approximately 0.87 mV K^{-1}. The non-isothermal temperature coefficient is a direct measure of the difference in potential of the two calomel electrodes at different temperatures, and is thus the more useful measure of the sensitivity of the electrode to changes in temperature.

The non-isothermal temperature coefficient of the saturated calomel electrode (SCE) is $+0.19 \text{ mV K}^{-1}$,[11] and of the 3.8 M calomel electrode $+0.44 \text{ mV K}^{-1}$.[10] A comprehensive resumé of temperature coefficient data is presented by Mattock.[10]

The standard potential including the liquid junction potential, as determined in cell 4, of the SCE is 0.2444 V at 25 °C and of the 3.5 M calomel electrode is 0.2501 V at 25 °C; these and further data are given by Bates.[8]

The methods of preparation of calomel electrodes have been well described by Hills and Ives.[15] Unless accurate and precise electrochemical measurements are to be made, in which the standard potential of the calomel electrode must be reproduced, the considerable labour in making a calomel electrode is not justified and a commercial electrode may be used. Commercial electrode assemblies usually have the electrode inverted, with the mercury uppermost, and packed into a narrow tube. A typical arrangement is shown in Fig. 2.1, together with a diagram of a silver/silver chloride reference electrode.

The stability of the potential of commercial calomel electrodes has been quoted by Mattock[10] as ± 0.1 mV over several months for electrodes stored in a thermostat near ambient temperature. Ferguson, Van Lente and Hitchens[16] found that the potentials of saturated calomel electrodes remained constant to ± 0.02 mV for several months, provided, as before, that they were stored in a thermostat. Such potential stability is essential if the cell potential is to be measured to ± 0.1 mV; thus temperature stability during measurements is very important.

The silver/silver chloride electrode

The silver/silver chloride electrode is in many respects the most satisfactory of all reference electrodes and certainly the simplest. In its rudimentary form the electrode consists of a silver wire or plate coated with silver chloride in contact with a solution of chloride ions saturated with silver chloride as shown in Fig. 2.1. It is commonly used as the internal reference electrode of pH electrodes and many other ion-selective electrodes.

A great advantage of the electrode is that if it is accidentally broken it does not release highly toxic chemicals as do the mercury-based electrodes and, especially, the thallium-based electrode. Thus the use of the silver/silver chloride electrode as against other electrodes is strongly to be recommended in applications where measurements are taken on or near food or drink.

The silver/silver chloride electrode is a reversible electrode of the second kind and its response, primarily to silver ions, is governed by the solubility equilibrium of the silver chloride. Thus chloride ions are sensed indirectly. For the same reasons as with the calomel electrode the source of chloride ions usually chosen is a strong potassium chloride solution.

The major problem with the electrode is the relatively high solubility of the silver chloride coating in the strong potassium chloride solution due to formation of polychloroargentate(I) complexes such as $AgCl_3{}^{2-}$. As the temperature rises, the solubility increases considerably; thus for use at high temperatures a sufficient excess of solid silver chloride must be added to the bridge solution to maintain saturation, otherwise silver chloride will dissolve off the electrode until saturation is reached, causing potential drift and shortening the life of the electrode.

Despite the apparent simplicity of the electrode, there is much discussion in the literature concerning the different methods of preparing the electrode. The aim of each method is to produce an electrode with the two components of the solid phase, the silver and the silver chloride, in their lowest energy states, and

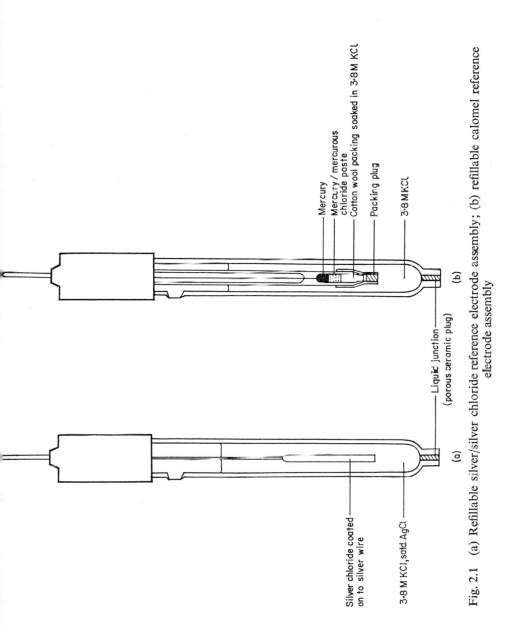

Mercury

Mercury/mercurous chloride paste

Cotton wool packing soaked in 3·8M KCl

Packing plug

3·8M KCl

(b)

Liquid junction (porous ceramic plug)

Silver chloride coated on to silver wire

3·8 M KCl, satd. AgCl

(a)

Fig. 2.1 (a) Refillable silver/silver chloride reference electrode assembly; (b) refillable calomel reference electrode assembly thus in the most reproducible or definable thermodynamic state. Most published methods involve the deposition of a layer of silver on a platinum substrate and subsequent formation of a chloridized layer; the two components of the solid phase should then be free of problems associated with mechanical strain, variations in crystal structure, impurities, nonstoichiometry, surface films and so forth.[17] A full discussion of these various methods is given by Janz;[17] electrodes so prepared are highly reproducible and stable to ± 0.02 mV or better.

However, for analytical work with ion-selective electrodes in cells with liquid junctions it is unnecessary to use such elaborate methods; precise knowledge of the potential of the electrode is not needed and adequate stability (better than ± 0.1 mV) is attained by electrodes prepared more simply. Commercial silver/silver chloride electrodes are usually prepared by electrolysis of pure silver wire or by dipping silver wire into molten silver chloride. The electrodes may be easily made in the laboratory if the following procedure is used.

(i) Clean a pure silver wire in strong ammonia solution; rinse; dip in 50% nitric acid until the silver is an even white colour (usually about one minute). Rinse with distilled water but do not dry.

(ii) Cathodize the wire against a silver anode at a current density of 0.5–10 mA cm^{-2} (about 5 mA cm^{-2} is most suitable) in 0.1 M hydrochloric acid (analytical reagent grade) for 30 s. Allow bubbles to disperse from the wire.

(iii) Anodize the wire for 10–15 min at the same current density. Rinse.

The electrode so prepared is dipped in a liquid junction tube containing a bridge solution of strong potassium chloride (3.8 M or 3.5 M is most convenient) saturated with excess silver chloride. The electrode will take several minutes, occasionally a few hours, to drift down to a stable potential. Freshly prepared electrodes are usually slightly positive (up to 5 mV) to aged electrodes; the ageing may be accelerated by heating in hot water at 60 °C for an hour.

The silver/silver chloride electrode is very stable over long periods in pure potassium chloride solutions but suffers interference, as with the calomel electrode, from redox reagents and species that react with silver chloride, such as sulfide, iodide, bromide and cyanide. Interference from such species in the sample is prevented by using a head of bridge solution across the liquid junction. Unlike the calomel electrode, however, there is a sufficiently large concentration of the electrode coating, silver chloride, dissolved in the bridge solution for problems still to arise particularly at high temperatures; interferents such as sulfide ions may react with the silver species coming through the junction and block the junction (with solid silver sulfide), thus causing increasing drift and instability of the electrode potential. Instability of the potential due to oxidation of silver chloride by atmospheric oxygen is not apparent unless an acidic bridge solution is used.

The effect of temperature on the electrode has already been partly considered in respect of the changing solubility of the silver chloride. The non-isothermal temperature coefficient of the electrode with a 4 M potassium chloride bridge solution is exceptionally low, $+0.09$ mV K^{-1},[11] and it is thus insensitive to small changes in temperature. The value of the non-isothermal temperature coefficient increases as the concentration of the potassium chloride bridge solution decreases, reaching $+0.75$ mV K^{-1} in 10^{-3} M potassium chloride solution.[10]

Thermal hysteresis effects are very small and usually negligible with silver/silver chloride reference electrodes. In experiments in the author's laboratory, 3.8 M silver chloride electrodes (using the same nomenclature abbreviation as for calomel electrodes) with dipped silver chloride elements were cycled from 25 °C to 100 °C four times and regained their initial potentials, to within 1 mV, rapidly after return to 25 °C. Of course, such treatment is extremely severe and unrealistic in the context of ion-selective electrode work. Silver/silver chloride electrodes have been used successfully up to 125 °C.[11]

Because of the relatively large exchange current at the silver chloride/solution interface and the large surface area of the electrode, the electrodes are relatively insensitive to polarization and recover quickly even after severe polarization.

The standard potentials, including the liquid junction potentials, as determined in a cell 4 arrangement, of the saturated and 3.5 M silver/silver chloride electrodes at 25 °C, are 0.1989 V and 0.2046 V, respectively; these and values at other temperatures are given by Bates.[8]

Finally, the effect of light on silver/silver chloride electrodes must be mentioned as it has been the subject of much discussion. Recent papers[19,20] have perpetuated the discussion. It may readily be appreciated from a survey of published results that they are inconclusive. In the author's opinion most of the reported effects of light may be attributed to temperature changes caused by the light rather than to genuine photoelectric effects. Investigations have been carried out in the author's laboratory in which the potential of 3.8 M silver/silver chloride reference electrodes (the element being of the dipped type) were measured against an identical electrode within a dark shield. The temperature difference between the electrodes was monitored by means of thermistors immersed in the bridge solutions. The unshielded electrode was subjected successively to sunlight, a 100 W tungsten filament light bulb at 30 cm and a 125 W mercury ultraviolet lamp at 10 cm. No potential change greater than 1.5 mV that could not be attributed to temperature effects was observed with aged electrodes, even with those drastically treated by the mercury lamp. Thus with such electrodes the effect of normal ambient light is small enough to ignore, unless high-precision experiments are to be performed in very variable lighting conditions.

The mercury/mercurous sulfate electrode

Although mercury/mercurous sulfate electrodes are commercially available and are widely used in applications where chloride contamination of the sample must be avoided, little information is available on them. Ives and Smith[21] have reviewed the literature and remark on the outstanding reproducibility of the electrode and its insensitivity to polarization. These properties have led to its use as half of the Weston Standard cell. They result from the relatively high solubility of the mercurous sulfate.

The electrode is similar in construction to the calomel electrode, and consists of a pool of mercury covered by a layer of mercurous sulfate in contact with a solution containing sulfate ions and saturated with mercurous sulfate. The mercurous sulfate is susceptible to hydrolysis, yielding a yellow basic salt; however, a small proportion of the basic salt in the solid layer does not seem to affect either the stability or the potential of the electrode.

Ives and Smith[21] give details of the various methods of preparing mercury/mercurous sulfate electrodes; the procedures are much less troublesome than those for preparing calomel electrodes. Most of the work reported on the electrode is of studies of cells using sulfuric acid as the source of sulfate ions. However, these systems are not useful as reference electrodes in ion-selective electrode work because of the high liquid junction potentials which would occur and the undesirability of contaminating samples with sulfuric acid. In reference electrodes strong sodium or potassium sulfate bridge solutions are normally used, although these still give larger junction potentials than potassium chloride bridge solutions.

The saturated potassium sulfate mercury/mercurous sulfate electrode has a potential of $+0.412$ V against SCE at 22 °C,[10] and the 1 M sodium sulfate electrode has a potential of $+0.402$ V against SCE at 22 °C.[22] The non-isothermal temperature coefficient of the 1 M sodium sulfate electrode is $+0.13$ mV K^{-1}.[22] The range of temperature in which the electrode can be used is 0–60 °C; as the temperature rises, the performance of the electrode is eventually limited by hydrolysis of the mercurous sulfate.

The thallium amalgam/thallous chloride electrode[18,23]

This electrode, which may be represented by the half-cell

$$\text{Tl, Hg (40\% m/m), TlCl(s)} \mid \text{KCl(sat)} \parallel$$

has been given the trade-name 'Thalamid' electrode by its manufacturers, Jenaer Glaswerk Schott und Gen. (Mainz). The principal advantages of this electrode are that it is remarkably free from thermal hysteresis and rapidly reaches equilibrium after temperature changes. The thallium amalgam is very sensitive to traces of oxygen in the bridging solution but fortunately the solubility of oxygen in saturated potassium chloride solution is very slight. The electrode may be used over the temperature range 0–135 °C. The standard potential of the electrode, including liquid junction potential, in pH 6.865 standard phosphate buffer was found by Baucke[23] to be -0.5767 V at 25 °C; standard potentials at other temperatures in the range 5–90 °C and in other buffers are also given by Baucke.

The thallium amalgam/thallous chloride electrode has the very serious disadvantage that accidental breakage of an electrode releases both compounds of thallium and the thallium amalgam, which are extremely dangerous because of their high toxicity (much worse than mercury compounds). The problem of safely disposing of a worn-out electrode is formidable.

Reference electrodes suitable for use in non-aqueous and partially aqueous systems have been reviewed by Hills[24] and, briefly, by Mattock.[10]

LIQUID JUNCTION POTENTIALS

At the junction between two electrolyte solutions, ions from the two solutions diffuse into each other. Different ions have different mobilities and will diffuse at different rates; the hydrogen ion, for example, is the most mobile ion and will diffuse faster than other ions. A charge separation will occur related in size to the difference in mobilities of the anions and cations in the two solutions. This charge separation produces a potential difference across the junction called the liquid junction potential.

In the situation described above, free diffusion of ions occurs in both directions across the junction and it is assumed that both solutions are static. In reference electrodes the situation is a little simpler than this. Usually the bridge solution is given a slightly higher pressure than the sample so that the solution, often strong potassium chloride solution, flows out relatively rapidly into the sample, and diffusion of the sample back into the salt bridge is impeded. If the bridge solution is made strong enough, it is assumed that variations in the liquid junction potential due to the varying composition of the sample are swamped.

Thus the liquid junction potential is, like the reference electrode potential, assumed to be constant, regardless of the composition of the sample in which the liquid junction and ion-selective electrode are immersed. This is the basis on which the reference electrode assembly is used. Any change in the liquid junction potential appears as a change in the potential of the assembly, since

$$E_{ref} = E_r + E_j$$

where E_{ref} is the potential of the assembly, E_r the potential of the reference electrode in the bridge solution and E_j the liquid junction potential. An extra liquid junction potential must be included if a double junction reference electrode is considered.

When an analysis using a cell with an ion-selective electrode is carried out, it is assumed that the liquid junction potential in the standard solutions used to calibrate the ion-selective electrode remains constant in all later measurements. Any change in liquid junction potential when the standard solutions are replaced by samples is termed the residual liquid junction potential[4] and constitutes an error in the analytical measurement. Thus the liquid junction potential must be as constant as possible. Constancy may be approached by a suitable choice of standards and/or sample pretreatment, and by use of the most satisfactory bridge solution and the best physical form of the liquid junction.

Where possible the complication of a second liquid junction in the cell should be avoided; occasionally, however, it is desirable, as discussed later, to use a double junction electrode in order to prevent contamination of a sample by the normal bridge solution. Substantial drift may occur unless this second junction is properly designed and looked after. In double junction reference electrodes the second junction (as shown in cell 3, p. 11) is between the bridge solution and the reference solution.

The liquid junction potential is affected by many parameters, the most important of which will be discussed in turn.

The junction structure

The choice of the type of junction to be used must be made from a range of devices ranging from the best in terms of stability and reproducibility, which are complicated to realize in practice, to the worst, which are easy to use but much less stable and reproducible.

Various types of liquid junction have been classified by Guggenheim[25] and additionally the flowing junction has been described by Lamb and Larson;[26] these classifications are discussed by Bates[8] and Mattock.[27] The junction usually used for the most precise work is the 'free diffusion' junction, in which the two liquids are brought together to form an initially sharp boundary with cylindrical symmetry. Alner, Greczek and Smeeth[28] have shown that such a junction is stable within 30–40 μV for several hours. This and the other well-defined junctions in the classification are impractical for analytical use as they are difficult to set up and maintain; most require large volumes of solution for running and some are more reproducible than is necessary in ion-selective electrode work.

Adequate stability and reproducibility can be obtained in nearly all applications with one of the commercial devices. These mostly fall into the category of 'restrained diffusion' junctions. In an assembly using such a junction a head of bridge solution maintains a slow but steady flow of the bridge solution into the sample through a restriction: the difference between the various types lies in the form of the restriction. The advantages and disadvantages of the types may be related to the stability of the electrolyte flow rate, the reproducibility of the flow rate and the ease with which the junction may be blocked.

The commonest of the commercial junctions available are the ceramic plug, asbestos wick or fibre, two types of ground sleeve junctions and the palladium annulus junction. These are shown in Fig. 2.2.

The first three have relatively slow flow rates (sometimes called leak rates) of about 0.01–0.1 ml per 5 cm head of bridge solution per day.[27] Ground sleeve junctions (type 1) have a flow rate of 1–2 ml per 5 cm head per day. The head of bridge solution is measured as the height of the surface of the bridge solution above the surface of the sample.

Leonard[29] reports that the flow rates of different asbestos wick junctions may vary by up to a factor of a hundred and that the liquid junction potential has a day-to-day stability of ± 2 mV under the favourable conditions of a junction between strong potassium chloride solution and an intermediate pH buffer. This may be compared to stabilities of ± 0.06 mV for the ground glass sleeve junction (type 1) and ± 0.2 mV for the now little-used palladium annulus junction in the same conditions. Mattock[27] has shown that the day-to-day reproducibilities of the ceramic plug and ground glass sleeve junction (type 1) are similar in intermediate pH samples but the ceramic plug is markedly better in samples of extreme pH (high or low). In a comparison of the ceramic plug with a capillary junction[30] both were found to be reproducible to ± 0.1 mV in aqueous buffers but the capillary junction was marginally worse in blood samples.

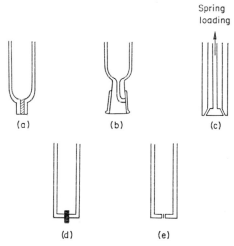

Fig. 2.2 Various forms of restrained diffusion liquid junction: (a) ceramic plug; (b) ground glass sleeve (type 1); (c) ground sleeve (type 2); (d) asbestos wick; (e) palladium annulus

Jackson[31] has pointed out that the palladium annulus junction partly responds as a redox electrode in strong oxidants (e.g. 0.2 M $KMnO_4$ in 0.05 M H_2SO_4) and mild or strong reductants (e.g. 0.5 M $SnCl_2$ in 1 M HCl). In such samples reference electrode assemblies with palladium or platinum annulus junctions should not be used.

The ground glass sleeve junction (type 1) has found particular use in applications where the junction has a tendency to clog, such as measurements in jam, protein, etc.; however, it has an inconveniently high flow rate necessitating frequent replenishment of the salt bridge. Asbestos wick junctions have been reported[14] to be particularly liable to blockage and consequently their use in samples other than clear solutions should be avoided.

No evaluation of the ground sleeve junction (type 2) has been reported. The method of reforming the junction is especially convenient and overcomes blockage problems; the reproducibility and stability of the junction have yet to be determined.

All these restricted junction devices require a head of bridge solution above the sample in order to work satisfactorily; thus the bridge should be kept topped up so that the surface of the bridge solution is always at least 1 cm above the sample solution. If at any time the bridge solution falls too low and remains so for long, the junction and bridge will become contaminated by species diffusing from the sample; the bridge solution should then be replaced. For the same reason, reference electrode assemblies should be stored when not in use with the junction immersed in bridge solution.

The mobility of sample species

Serious errors may be introduced due to residual liquid junction potentials if the sample has an extreme pH value, either high or low. Hydrogen ions are nearly five times and hydroxyl ions nearly three times more mobile than potassium or chloride ions, whereas most other ions have mobilities within $\pm 30\%$ of potassium and chloride ions. Thus changes in the activity of hydrogen or hydroxyl ions in a sample have a much greater effect on the liquid junction potential than changes in other ionic activities; when the pH of the sample lies outside the range 2–12, these effects are significant.

The problem of measurements in samples of extreme pH with an ion-selective electrode may usually be solved by buffering the sample to a neutral pH in a pretreatment stage. Where such pretreatment is not possible (e.g. to prevent upsetting chemical equilibria) and the concentration of acidic or basic constituents in the sample remains constant, the salt bridge may be changed to match the sample. If the sample background cannot be used as a bridge solution with any of the available reference electrodes, then a double junction reference electrode assembly will have to be used. In theory the only requirement in these measurements at extreme pH is to match the standards to the samples closely (thus again making the assumption that the sample composition remains constant) and thus reduce the residual liquid junction potential to zero. However, the large liquid junction potential which would be obtained if neither the samples were treated nor the bridge solution modified would be insufficiently stable or reproducible for this residual potential to be negligible. As an example of the size of the potentials which can be produced, the liquid junction potential across the junction 1 M HCl ‖ KCl (sat) is approximately 14 mV, and across the junction 1 M NaOH ‖ KCl(sat) is −6.9 mV, whereas between a saturated potassium chloride solution and a buffer of intermediate pH it is only approximately 3 mV.[4]

Baumann[32] has studied this pH effect in an experiment with a series of solutions of constant ionic strength but varying perchloric acid concentration. When a silver/silver chloride reference electrode with the unusual 2 M sodium chloride bridge solution and a sample of the same ionic strength were used, a change in acid concentration from 0 to 0.04 M produced a change in liquid junction potential of 1.1 mV and from 0 to 0.48 M a change of 12.3 mV. The work of Biedermann and Sillén gave comparable results.[33]

The bridge solution

The bridge solution, often called the salt bridge, should be selected to minimize the liquid junction potential and also to minimize contamination of the sample by, for example, extra determinand or species which can interfere with the response of the ion-selective electrode. The junction potential may be minimized either by using an equitransferent solution[34] (i.e. a solution in which the total mobilities of all the ions in the solution may be summed such that

$$\sum_i n_i a_i u_i = 0$$

where n_i, a_i and u_i are the charge, activity and mobility of ion i) or by using a solution virtually identical with the sample. The former approach is more usual since a reference electrode assembly containing such a bridge solution is of general application and fulfils the requirement that the ionic strength of the bridge solution should be much (preferably more than ten times) greater than that of the sample. The latter approach is, however, very useful in applications where the composition of the sample remains approximately constant because the junction potential is reduced to virtually zero. This method was mentioned in the previous section in relation to acidic samples but may also be used for samples with a very high ionic strength.

Strong potassium chloride solutions are the most widely used bridge solutions. The strength of the potassium chloride solution is usually saturated, or 3.8 M or 3.5 M. The saturated solution (4.16 M at 25 °C) is the easiest to prepare and its use in calomel electrodes as bridge solution (cell 2) or reference solution (cell 3) is advantageous as the SCE has the lowest non-isothermal temperature coefficient of all calomel electrodes. Saturated potassium chloride solution is the only solution used in thallium amalgam/thallous chloride electrodes. However, in electrodes using this saturated bridge solution accumulation of crystals inside the assembly, round the liquid junction, tends to make the junction irreproducible and unstable by partial or complete blockage of the junction. Blockage is also particularly likely if the junction is rapidly cooled—for example, by transference from a hot solution to a much cooler solution—as potassium chloride then crystallizes out within the junction.

Picknett[35] has shown, both theoretically and practically, that the liquid junction potential between saturated potassium chloride and other solutions increases as the concentration and, hence, the specific conductivity of the sample solution decreases. A summary of some of his results is given in Table 2.1. With an ideal bridge solution the liquid junction potential would decrease towards zero as the sample became more dilute.

TABLE 2.1
Calculated liquid junction potentials between various solutions and saturated KCl solution at 25 °C mV^{-1} (from Ref. 35)

Concentration/M	HCl	KCl	KOH	KH phthalate	CH$_3$COONa/CH$_3$COOH
10^{-1}				2.8	2.2
10^{-2}	2.9	2.8	1.9	3.5	3.2
10^{-3}	4.0	3.9	3.2	4.1	4.2
10^{-4}	4.8	5.0	4.5	4.9	5.0
10^{-5}	5.7	6.1	5.8	5.8	5.8
10^{-6}	6.7	7.1	6.9	6.7	6.7

Leyendekkers,[36] using reference electrodes with saturated potassium chloride bridge solutions, showed that the liquid junction potential varied linearly with the ionic strength of strong sample solutions of pure sodium or potassium chloride. The junction potentials were approximately twice as large in the sodium chloride solutions as in the potassium chloride solutions at the same ionic strength.

The 3.8 M solution is often chosen as an alternative as it does not crystallize out until cooled to 0 °C, and thus in normal use the errors associated with the presence of the crystals are avoided. Bates,[4] who recommends the 3.5 M solution, reports that he has been unable to detect errors, due to changes in junction potential, which were attributable to the concentration of potassium chloride bridge solutions when this was in the range 3 M to saturated and the temperature was between 10 and 40 °C.

The ubiquitous use of strong potassium chloride solutions as bridge solutions is evidence of their reliability in pH measurement; however, in such measurements changes in junction potential of a few tenths of a millivolt frequently pass unnoticed (± 0.01 pH unit $\rightarrow \pm 0.6$ mV). In some measurements with ion-selective electrodes smaller junction potentials would be advantageous. Therefore it is worth while to consider the use of alternative bridge solutions, some of which have been formulated to be more closely equitransferent than potassium chloride solution. Unfortunately there is not yet much evaluation data on these solutions. Four compositions are given in Table 2.2.

TABLE 2.2
Equitransferent bridge solutions

Source	Composition
1. Grove-Rasmussen[37]	1.8 M KCl + 1.8 M KNO$_3$
2. Kline, Meacham and Acree[38]	3 M KCl + 1 M KNO$_3$
3. Orion Res. Inc. Type 90-00-01	1.70 M KNO$_3$ + 0.64 M KCl + 0.06 M NaCl + 1 ml/l of 37% HCHO
4. Wilson, Haikala and Kivalo[39]	2 M RbCl

For a wide variety of applications it might reasonably be expected that any of the listed compositions would give somewhat lower and more reproducible and stable junction potentials than potassium chloride solution, but the improvement would not be very large. Johansson and Edström[40] compared 1 M NaNO$_3$, 1 M KNO$_3$, saturated KCl and the Orion formulation, as bridge solutions for a reference electrode contacting dilute copper nitrate solutions. The saturated KCl solution and the Orion formulation gave similar potentials in most samples, with the Orion solution being marginally better where differences were apparent; the other bridge solutions gave markedly larger liquid junction potentials.

When one of the equitransferent solutions or a potassium chloride solution is used as bridge solution in a single junction silver/silver chloride reference electrode assembly, a little silver nitrate should be added to saturate the solution with silver chloride; otherwise the electrode readings will drift initially until saturation has been achieved by dissolution of silver chloride from the reference electrode. Thus the Orion filling solution type 90-00-02, which is in direct contact with the AgCl pellet as bridge solution in single junction reference electrodes or as the inner solution in double junction reference electrodes, is identical with 90-00-01 except that it is saturated with silver chloride.

As previously mentioned, the bridge solution is chosen so that the leakage of this solution into the sample results in the minimum contamination of the sample, either by determinand ions or by interferents. Thus it is clearly unsatisfactory to use a bridge solution of strong potassium chloride if measurements are to be made on chloride samples with very low concentrations, since chloride ions leaking through the junction would lead to erroneously large answers. Similarly, the same bridge solution would be unsuitable both in the measurement of very dilute ammonium samples with the ammonium-selective electrode, because potassium ions interfere with the ammonium electrode, and in the measurement of dilute nitrate samples with the nitrate electrode, as then chloride ions would interfere with this electrode. In such cases the best course is to use either a double junction electrode with an innocuous bridge solution or to use a different reference electrode altogether. In the first example above, either a double junction calomel or silver/silver chloride reference electrode, with a potassium nitrate bridge solution, could have been used or else a mercury/mercurous sulfate reference electrode with a 1 M sodium sulfate bridge solution. In the other examples interference is usually reduced sufficiently by using a 0.1 M potassium chloride bridge solution: the choice of bridge solution for use with the nitrate electrode is discussed more fully in Chapter 6. To overcome interference from potassium chloride bridge solutions in potassium measurements, a lithium trichloroacetate bridge solution in a silver/silver chloride single junction reference electrode has been recommended.[41] In all these cases contamination could be avoided by carrying out the measurements in a flowing system with the reference electrode assembly sited downstream from the ion-selective electrode; thus, while the system is operating, no contaminants can reach the ion-selective electrode.

Although it is necessary to be aware of the danger of contamination from bridge solutions, the amount of contamination which can occur during the few minutes needed for most measurements will often be insignificant. For example, if a saturated potassium chloride bridge solution leaks from a head of 5 cm through a liquid junction device with a flow rate of 0.1 ml per 5 cm head per day into a 50 ml sample for 5 min, the increase in the concentration of potassium and chloride ions in the sample will be about 3×10^{-5} M. If this value is small with respect to the sample concentration, then the contamination may be ignored; if not, then the alternatives are either to change the bridge solution or to choose a liquid junction with a slower flow rate. This contamination problem is the more serious the longer the reference electrode assembly is left in the sample; thus special care is necessary, for example, when the electrodes are being used to follow a potentiometric titration or when the response time of the ion-selective electrode is slow.

A bridge solution which reacts with the sample is also unsatisfactory. Thus perchlorate ions, added perhaps as a constant ionic strength background, will react with potassium ions leaking through a junction to form a precipitate of potassium perchlorate which may block, partially or completely, the junction, causing rising and unstable junction potentials. In such a case a mercurous sulfate electrode with a 1 M sodium sulfate bridge solution is a suitable alternative. As a second example, single junction silver/silver chloride reference electrodes are not useful in samples containing sulfide ions, since the sulfide ions will react with ionic silver species in the bridge solution and may cause blockage of the junction with silver sulfide; calomel electrodes are usually better in this respect as the concentration of dissolved mercury species is very much smaller than that of the silver species in silver/silver chloride electrodes. However, the use of a double junction electrode with a strong potassium chloride bridge solution is possibly the most satisfactory answer for samples containing sulfide ions.

The choice of bridge solution for measurements in samples of extreme pH has been considered in the previous section. The use of a cell without a liquid junction, if a suitable reference system can be found, is advantageous in these circumstances.

Colloid and suspension effects

In the same way that colloids and suspensions affect the liquid junction potential in pH measurement,[8] they will affect liquid junction potentials in ion-selective electrode work. A liquid junction immersed in a slurry of charged particles— for example, a slurry of soil or solid ion exchanger—may develop a potential different by as much as 240 mV, or even more, from an identical junction in the supernatant liquid.[42,43] An attempt may be made to overcome this source of error (the large residual liquid junction potential) by dispersion of colloids before measurement or by taking measurements only in the supernatant liquid of suspensions.

THE CONSTRUCTION OF REFERENCE ELECTRODE ASSEMBLIES

Reference electrode assemblies may be constructed in several ways. The different configurations are obtained by permuting the following alternatives:

(i) The assembly may be of single or double junction pattern (as in cells 2 and 3, respectively).

(ii) The solution compartments may be refillable or sealed.

(iii) The assembly may be separate or incorporated into the ion-selective electrode assembly to make a dual electrode (sometimes called a 'combination electrode').

Some of the commonest permutations are shown in Figs 2.1 and 2.3.

As discussed in the section on bridge solutions, occasionally it is advantageous to use a double junction assembly, particularly if contamination of the sample is possible and the electrodes are to be left in the sample for a long time. The inner junction (J_1 in cell 3) of a double junction assembly needs as much careful maintenance as the outer junction, and both the reference solution and bridge solution should be renewed regularly, at least once a week; in assemblies where the reference solution is not renewable, potential drift will inevitably occur as reference solution becomes mixed with bridge solution. This disadvantage is aggravated in some commercial electrodes by having only a small volume of reference solution. In a double junction reference electrode, the total liquid junction potential ($E_{J_1} + E_{J_2}$) is largely independent of the composition of the bridge solution, and has a value similar to that of a junction between the reference solution and the sample in a single junction assembly. Thus, use of a double junction assembly usually does nothing to minimize the liquid junction potential but increases the uncertainty concerning potential stability of the system.

Dual electrode assemblies, comprising an ion-selective electrode and a reference electrode in one probe, are very convenient to use, as only one 'electrode' need be put into the sample. Unfortunately only a few such 'electrodes' are commercially available. Despite the potential advantages of these 'electrodes', it has been found that some of those which are available perform less satisfactorily than the same electrodes separately, often exhibiting poor stability; this is due to poor mechanical design of the particular 'electrodes' (e.g. poor liquid junctions or insufficient insulation in the cap between the connections to the two electrodes) rather than to a fault inherent in the concept of dual electrodes. Dual pH and pNa electrodes have been manufactured for several years and are generally just as satisfactory as the separate electrode systems.

Sealed reference electrode assemblies are particularly common in those industrial applications in which infrequent maintenance is essential and poorer accuracy can be tolerated. A typical sealed assembly has a single liquid junction and a large bridge solution container packed with a thick slurry of potassium chloride (gelled solutions are also used): the reference electrode is nearly always

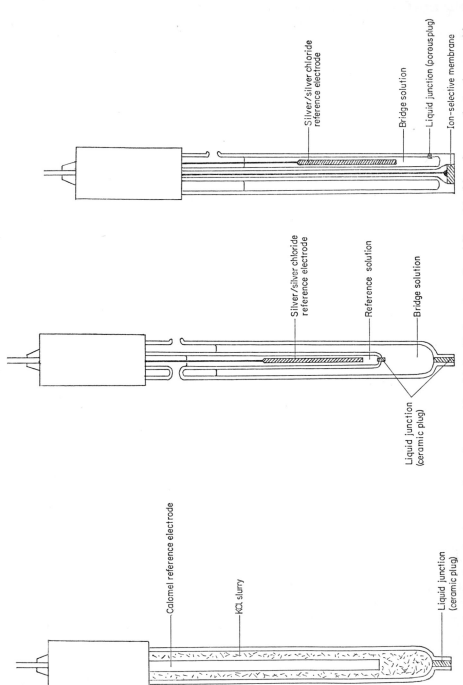

Fig. 2.3 (a) Sealed calomel reference electrode assembly; (b) double junction, silver/silver chloride reference electrode assembly; (c) dual electrode, incorporating an ion-selective electrode based on an inorganic salt and a silver/silver chloride reference electrode

a calomel or silver/silver chloride electrode. The potassium chloride from the slurry leaks slowly out of the electrode, making the bridge solution progressively more dilute; eventually, when drift becomes significant, the whole assembly is discarded.

A sealed assembly is more prone to contamination than others because of the slow flow rate of the junction; this slow flow rate also gives rise, in some circumstances, to larger and more irreproducible junction potentials.

Temperature fluctuations also create problems. If the temperature of the sealed electrode assembly is raised, pressure builds up inside and forces out the bridge solution more rapidly than usual; however, if the temperature is substantially reduced, the pressure inside the electrode drops and samples plus contaminants are sucked inside the electrode. Thus, temperature fluctuations lead to rapid contamination and exhaustion of the electrode. Another danger is that if the assembly is left with the junction out of the solution for very long, because of the relatively small amount of liquid in the bridge the whole assembly will dry out and must be discarded. Like all reference electrode assemblies, sealed assemblies perform best if stored with the junction soaking in bridge solution.

REFERENCES

1. S. E. Manahan, *Anal. Chem.* **42**, 128 (1970).
2. J. Mertens and D. L. Massart, *Bull. Soc. Chim. Belg.* **82**, 179 (1973).
3. K. M. Stelting and S. E. Manahan, *Anal. Chem.* **46**, 2118 (1974).
4. R. G. Bates, *Determination of pH, Theory and Practice*, 2nd ed., Chapter 3, Wiley, New York (1973).
5. Orion Research Inc., *Newsletter* **5**, 1 (1973).
6. G. J. Janz and D. J. G. Ives, *Ann. N. Y. Acad. Sci.* **148**, 210 (1968).
7. D. J. G. Ives and G. J. Janz (eds.), *Reference Electrodes*, Academic Press, New York (1961).
8. R. G. Bates, *loc. cit.*, Chapter 10.
9. A. K. Covington, in *Ion-Selective Electrodes*, R. A. Durst, editor, Chapter 4. N.B.S. Spec. Publ. No. 314, Washington, D.C. (1969).
10. G. Mattock, *pH Measurement and Titration*, Chapter 7, Heywood, London (1961).
11. J. E. Leonard, in *Symposium on pH Measurement*, p. 16, A.S.T.M. Spec. Tech. Pub. No. 190, Philadelphia (1957).
12. L. Van den Berg, *Anal. Chem.* **32**, 628 (1960).
13. G. D. Pinching and R. G. Bates, *J. Res. Nat. Bur. Stand.* **37**, 311 (1946).
14. G. Mattock, *loc. cit.*, Chapter 11.
15. G. J. Hills and D. J. G. Ives, Chapter 3 in Ref. 7.
16. A. L. Ferguson, K. Van Lente and R. Hitchens, *J. Am. Chem. Soc.* **54**, 1279 (1932).
17. G. J. Janz, Chapter 4 in Ref. 7.
18. H. K. Fricke, *Zucher* **14**, 3 (1961).
19. G. J. Moody, R. B. Oke and J. D. R. Thomas, *Analyst* **94**, 803 (1969).
20. R. A. McAllister and R. Campbell, *Anal. Biochem.* **33**, 200 (1970).
21. D. J. G. Ives and F. R. Smith, Chapter 8 in Ref. 7.
22. P. L. Bailey, unpublished work in the E.I.L. laboratory.
23. F. G. K. Baucke, *J. Electroanal Chem.* **33**, 135 (1971).
24. G. J. Hills, Chapter 10 in Ref. 7.

25. E. A. Guggenheim, *J. Am. Chem. Soc.* **52**, 1315 (1930).
26. A. B. Lamb and A. T. Larson, *J. Am. Chem. Soc.* **42**, 229 (1920).
27. G. Mattock, *loc. cit.*, Chapter 8.
28. D. J. Alner, J. J. Greczek and A. G. Smeeth, *J. Chem. Soc.* (*A*), 1205 (1967).
29. J. E. Leonard, reported in Ref. 27.
30. G. Mattock and D. M. Band, in *Glass Electrodes for Hydrogen and other Cations*, G. Eisenman, editor, Chapter 2. Marcel Dekker, New York (1967).
31. J. Jackson, *Chem. Ind.* 272 (1969).
32. E. Baumann, *J. Electroanal Chem.* **34**, 238 (1972).
33. G. Biedermann and L. G. Sillén, *Ark. Kemi* **5**, 425 (1953).
34. K. V. Grove-Rasmussen, *Acta Chem. Scand.* **2**, 937 (1948).
35. R. G. Picknett, *Trans. Faraday Soc.* **64**, 1059 (1968).
36. J. V. Leyendekkers, *Anal. Chem.* **43**, 1835 (1971).
37. K. V. Grove-Rasmussen, *Acta Chem. Scand.* **3**, 445 (1949); **5**, 422 (1951).
38. G. M. Kline, M. R. Meacham and S. F. Acree, *Bur. Stand. J. Res.* **8**, 101 (1932).
39. M. F. Wilson, E. Haikala and P. Kivalo, *Anal. Chim. Acta* **74**, 395, 411 (1975).
40. G. Johansson and K. Edström, *Talanta* **19**, 1623 (1972).
41. Orion Research Inc., *Analytical Methods Guide*, 7th edn, Cambridge, Mass. (1975).
42. H. Jenny, T. R. Nielsen, N. T. Coleman and D. E. Williams, *Science, N.Y.* **112**, 164 (1950).
43. I. Tasaki and I. Singer, *Ann. N.Y. Acad. Sci.* **145**, 36 (1968).

CHAPTER 3

ELECTRODE PERFORMANCE

In this chapter some of the essential theories and features of the performance of ion-selective electrodes and gas-sensing probes will be introduced in general terms. Detailed considerations of these points in respect of particular sensors will follow in the relevant chapter. In addition, a set of points for assessing the performance and utility of electrodes and probes is given. The discussion highlights those factors which must be considered in comparing different types of electrode or probe which sense the same species.

GENERAL THEORY

Ion-selective electrodes sense the activities of ions in solutions; if the environment of the ions is known and well defined, the activities may be related to the concentrations of the ions.

The activity, a_X, of an ion X in solution is related to its concentration, m_X, expressed as a molality (moles of X per kilogram of solvent), by the equation

$$a_X = \gamma_X m_X \tag{3.1}$$

where γ_X is the activity coefficient. As will be discussed later, it is not possible to determine the activity coefficient of any one type of ion in solution in isolation from the others. Thus, in a solution of potassium chloride, it is not possible to measure the individual activity coefficients of the potassium and chloride ions, γ_{K^+} and γ_{Cl^-}, but only a mean activity coefficient, $\bar{\gamma}^2_{KCl}$, which is defined by

$$\bar{\gamma}^2_{KCl} = \gamma_{K^+} \gamma_{Cl^-} \tag{3.2}$$

In the general case of the ion X in solution with a counter ion Y, the mean activity coefficient, $\bar{\gamma}_{XY}$, is dependent upon the ionic strength of the solution.

According to Debye–Hückel theory, this dependence may be expressed in the form

$$\log \bar{\gamma}_{XY} = \frac{-A|n_X n_Y|I^{\frac{1}{2}}}{1 + BdI^{\frac{1}{2}}} \qquad (3.3)$$

where A and B are conditional constants depending upon such variables as temperature and the density and dielectric constant of the solvent; n_X and n_Y are the charges on ions X and Y; d is termed the average effective diameter of the ions; and I, the ionic strength, is defined by

$$I = \tfrac{1}{2} \sum m_X n_X{}^2 \qquad (3.4)$$

where the summation is taken over all the ions in the solution.

In analytical work it is usual to measure concentrations on the molar scale ([X], moles of X per litre of solution) or in related units (milligrams of X per litre of solution, i.e. ppm). Concentrations on the molal and molar scales are virtually identical except in very strong solutions of X; thus, a_X may be expressed by the equation

$$a_X = \gamma_X [X] \qquad (3.5)$$

without significant error in most analytical work.

An ideal ion-selective electrode for ion X produces a potential E in solutions of this ion given by the Nernst equation

$$E = E^\circ + \frac{2.303 RT}{n_X F} \log_{10} a_X \qquad (3.6)$$

where E° is the standard potential of the electrode and R, T and F have their conventional meanings. Thus, the electrode potential is proportional to the logarithm of the activity of ion X. The potential will also be proportional to the logarithm of the concentration if the activity coefficient remains constant; in this case Eq. (3.6) reduces to

$$E = E^{\circ\prime} + \frac{59.12}{n_X} \log_{10} [X] \quad \text{at } 25\,^\circ C \qquad (3.7)$$

For pure solutions of XY, of concentration less than about 10^{-4} M, the activity coefficient is close to unity, and differences between activity and concentration may be ignored. However, as the concentration increases above 10^{-4} M the activity coefficient decreases and these differences become progressively larger. Thus, a calibration graph, produced from electrode measurements in concentration standards prepared by serial dilution, will be curved at the higher concentrations, as shown in Fig. 3.1. When the standards are treated by addition of a high concentration of inert electrolyte to produce a constant ionic strength background, the activity coefficient is stabilized and a straight line calibration graph is obtained, allowing concentrations to be measured directly.

If the determinand X is partially complexed in the sample, as are, for example, fluoride ions in drinking water or hydrogen ions in a phthalate buffer, the electrode will still sense the activity of the uncomplexed or 'free' ions. As the determinand is not consumed by the electrode, the complexation equilibria are not displaced and, hence, ion-selective electrodes offer a valuable method for determining the uncomplexed ion activity.

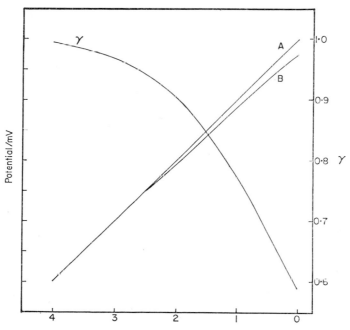

Fig. 3.1   Calibration curves (left-hand ordinate) produced from activity standards, $-\log_{10} a_X$ (A) and concentration standards, $-\log_{10} c_X$ (B), shown with the corresponding plot of the activity coefficient, γ (right-hand ordinate)

In most analytical techniques, such as titrimetry, spectrophotometry or flame photometry, the 'total' concentration of both complexed and uncomplexed determinand species is measured. Such measurements are also possible with ion-selective electrodes if the sample is pretreated to fix the activity coefficient and decomplex the determinand. The important choice of the correct pre-treatment solution is considered in the section on concentration standards and in Chapter 9.

Thus, ion-selective electrode techniques are very flexible. Variation of the reagent to pretreat the sample allows a range of different determinands (i.e. different forms of the species sensed by the electrode) to be measured; therefore careful choice of calibration standards for the electrode is essential.

CALIBRATION OF ELECTRODES

An ion-selective electrode can be calibrated either with solutions of known determinand activity or with solutions of known determinand concentration, according to which parameter is to be measured in the samples.

Activity standards

To prepare a standard solution containing a known activity of any ion is strictly impossible, as it requires knowledge of the activity coefficient of that ion. On theoretical and practical grounds such single ion activity coefficients cannot be determined without recourse to a non-theoretical, semi-arbitrary assumption. In the case of the solutions of 'known' hydrogen ion activity used to establish the pH scale, two almost consistent assumptions were made for the National Bureau of Standards (US) scale and the British Standards Institution scale. The NBS chose to ascribe a conventional value to the chloride ion activity coefficient in dilute solutions (the Bates–Guggenheim convention), whereas the BSI took the direct approach and ascribed a conventional value to the hydrogen ion activity in a 0.05 M potassium hydrogen phthalate solution. In the formulation of series of solutions of standard activities of other ions, a similar assumption or convention is explicitly or implicitly used.

The choice of conventions available for the assignment of activities to solutions for the calibration of ion-selective electrodes has been discussed by Bates.[1] When the ionic strengths of solutions are below 0.1, the straightforward MacInnes convention ($\gamma_{Cl^-} = \gamma_{K^+}$ in KCl solutions) may be used, or the Debye–Hückel convention (based on Eq. (3.3), which gives $\gamma_X = \gamma_Y$ for any univalent electrolyte XY) or the pH convention (fixing a value for γ_{Cl^-} from Eq. (3.3), with conventional values for A and Bd). If the convention only gives the activity coefficient of one of the ions in the simple electrolyte, the activity coefficient of the other may be calculated from a general form of Eq. (3.2), i.e.

$$\bar{\gamma}_{XY}^2 = \gamma_X \gamma_Y \tag{3.8}$$

where $\bar{\gamma}_{XY}$ is the measurable mean activity coefficient, again for a univalent electrolyte XY. Additionally, the activity coefficient may be assumed to have the same value in other electrolytes of the same charge type at the same ionic strength; hence, by further use of Eq. (3.8) more activity coefficients may be obtained.

For example, if the activity coefficients γ_{K^+} and γ_{Cl^-} are 'known' from measurements of $\bar{\gamma}_{KCl}$ in a given solution of potassium chloride and adoption of the MacInnes convention, the activity coefficient of sodium ions, γ_{Na^+}, in solutions of sodium chloride of the same ionic strength, can be calculated from measured values of $\bar{\gamma}_{NaCl}$ by means of the equation

$$\gamma_{Na^+} = \bar{\gamma}^2_{NaCl}/\gamma_{Cl^-} \tag{3.9}$$

This procedure may then be extended using this value of γ_{Na^+} to calculate, for example, γ_{Br^-} in solutions of sodium bromide from measured values of γ_{NaBr}.

Alternatively, if the convention is consistent, the same value of γ_{Br^-} should be obtainable by use of the original value of γ_{K^+} and measurements of the mean activity coefficient, $\bar{\gamma}_{KBr}$. Bates[1] has compared activities calculated from each convention for a range of electrolytes, including some containing polyvalent ions, and has shown that the choice of convention has relatively little effect.

However, when the ionic strengths of solutions are greater than 0.1, a more complex theory is required to account for the different properties of the different ions and also for the dependence of the ionic activities upon the composition of the solution, instead of only upon the ionic strength. The concept of ionic strength becomes progressively less useful as the concentrations increase. Bates, Staples and Robinson[2] have proposed a convention for these concentrated solutions which is based on the hydration theory of Stokes and Robinson.[3] This theory seeks to explain ionic activities in terms of the number of water molecules associated with each ion (a measure of which is referred to as the hydration number) and the activity of the residual unassociated water. The convention adopted within the terms of this theory is the assignment of a hydration number to the chloride ion; values of 0 and 0.9 have been proposed. The uncertainty in these hydration numbers has been assessed by Bates[1] as 0.5–1.0. After one value is assumed, all the rest may be calculated from the experimental results. A full discussion of this theory is given in the book by Robinson and Stokes.[4]

At the other extreme of concentration, for solutions of ionic strength less than 10^{-4}, the activity coefficient may be assumed to be unity, making the activities of ions equal to their concentrations.

Several ranges of activity standards have been proposed, employing one or other of these conventions; they are mostly standards for alkali metal and halide ions. A summary of solutions proposed as standards and others which may be used for calibration purposes is given in Table 3.1: standards developed especially

TABLE 3.1
Solutions of measured or calculated activity

| mol kg^{-1} | Salt | | | |
	KF $pF = pK$	CaCl$_2$ pCa	NH$_4$Cl pNH$_4$	KCl pCl
0.01	2.044			2.044
0.0333		1.900		
0.05	1.387			
0.1	1.111	1.570	1.112	1.113
0.2	0.837	1.349	0.840	0.840
0.5	0.475	0.991	0.483	0.486
1.0	0.190	0.580	0.208	0.216
2.0	−0.119	−0.198	−0.080	−0.059

Data adapted from Refs 2, 5 and 6. $pX = -\log_{10} a_X$.

for the single-point calibration of electrodes for the analysis of calcium and potassium in blood serum[7] are detailed in Table 3.2.

Once a standard activity scale has been set up for an ion and a range of activity standards specified, these activity standards may be used for calibrating ion-selective electrodes, which may then be used for the direct determination of ionic activities in samples. Such measurements are difficult or impossible by other techniques and, hence, this usage of ion-selective electrodes is particularly important for both the analyst and the physical chemist.

TABLE 3.2
Reference standards for the determination of pK and pCa in blood serum[7]

| Solution composition | | | Standard values | | |
NaCl	KCl	CaCl$_2$	pNa	pK	pCa
0.1450 M	0.0042 M			2.504	
0.1454 M		0.001 26 M			3.360
0.1414 M	0.0041 M	0.001 21 M	0.966	2.514	3.373

For all three solutions the ionic strength is 0.15 mol kg^{-1}.

The direct and rapid determination of ionic activities is useful in both analytical applications and fundamental studies of solution chemistry. The determination of calcium activity in blood by use of a calcium selective electrode is an example of an analytical application in which ionic activity rather than concentration is the parameter which must be measured and which is difficult by any other technique. Fundamental work in which activity measurements with ion-selective electrodes have been used includes studies of the states of ions in concentrated solutions and experiments on ionic complexation: examples are given in later chapters.

Concentration standards

The analyst usually wishes, rightly or wrongly, to measure the concentrations of ions rather than their activities and thus concentration standards are far more widely used than activity standards. Moreover, concentration standards may readily be prepared for any ion without the relatively difficult problems associated with single ion activities discussed in the previous section. No conventional scales need to be established; it is simply sufficient to prepare a range of solutions of the required concentrations and take steps to ensure the constancy of the activity coefficient. The ion-selective electrode may then be calibrated in these concentration standards and used to determine the concentration of the determinand in other solutions in which the activity coefficient of the determinand is the same as in the standards. The ionic activity, which the electrode senses, is then proportional to the ionic concentration (Eqs. (3.1, 3.6 and 3.7)).

The purpose of the calibration is to enable the response of the electrodes in standard solutions to be compared with the response in samples. In order for the comparison to be valid, both standards and samples must be treated identically. Thus, any reagents added to samples before measurement should also be added to standards in the same proportions, so that the background compositions of samples and standards are identical. If sample pretreatment is not necessary, because the conditions for satisfactory measurement are already met by the sample in its untreated state, then it is necessary to prepare the standards with the same background composition as the sample. Concentration standards as presented to the electrode should thus be as similar as possible in all respects to the samples, and the determinand concentration in the standards should closely bracket the expected range in the samples.

When a buffer system is used for calibrating the electrode, the standards are not usually very similar in composition to the samples, and interpretation of the potential measurements in samples becomes more difficult. Although the ionic strength of both samples and standards may be similar and, hence, nominally the activity coefficients are the same, the immediate ionic environments of the ions in the standards and samples are substantially different. Hence, the desired correlation between activity and concentration may not exist and it is better to regard the measurement as one of activity at constant ionic strength.

The range of concentration standards for calibrating an electrode may usually be prepared by serial dilution of a standard stock solution of a salt containing the determinand. Normal criteria of analytical chemistry should be used in the selection of this salt so that, where possible, the salt is of a high and defined purity and the stock solution is stable for long periods to avoid the necessity for frequent standardization. The more dilute standards, particularly those containing less than 10^{-3} M determinand, should be prepared by dilution of the stock solution immediately before use because of their instability. In some cases—for example, solutions of sulfide ions—soluble salts are not available in sufficiently pure form and moreover the solutions are unstable; thus, the stock standard solution must always be standardized before use. Similar to the problem with sulfide standards is that with sulfur dioxide standards; no salts are available with sufficiently precise purity for calibrating a sulfur dioxide probe. The purity of most salts, such as potassium metabisulfite or sodium sulfite, is usually quoted as a minimum of about 96%. Additionally, the solutions are subject to aerial oxidation and must therefore be standardized by an iodine-thiosulfate titration method. However, in most cases, because of the comparatively limited accuracy and precision of ion-selective electrode techniques, selection of a suitable salt is not difficult, especially in comparison with the difficulty of selecting standards for techniques of high accuracy and precision such as coulometry or even volumetric titrimetry. Coulometric generation[8] of low concentration standards has been used to increase accuracy and reduce solution handling: unfortunately only a few ions, notably the halide ions and silver ions, may be generated in this way.

In conjunction with the concentration standards, a reagent must be used to stabilize the activity coefficient of the determinand and perform other important functions such as buffering the pH of the sample and decomplexing the determinand. The formulation of this reagent will depend on the chemical properties and concentration of the determinand, the form in which it is to be measured and the optimum working conditions for the electrode. Commonly, samples have low ionic strengths and the determinand concentration is also low ($<10^{-2}$ M); in such cases it is best to limit the concentration of the background electrolyte, in the treated samples and standards, to about 10^{-1} M to minimize difficulties with the liquid junction potential and minimize sample dilution.

A well-known example of a reagent formulated to meet all these conditions is that called T.I.S.A.B. (Total Ionic Strength Adjustment Buffer), which is added to drinking water samples before analysis for fluoride ions with the fluoride electrode. The reagent contains sufficient sodium chloride to give a background concentration of 0.1 M in the treated sample and, hence, fix γ_{F^-}, acetate buffer to buffer the sample to the optimum working pH (about pH 5–6) and 1,2-diaminocyclohexane-N,N,N',N'-tetra-acetic acid (CDTA) to complex such metal ions as aluminium and ferric which otherwise complex fluoride ions. Relatively large changes in the background of the untreated samples do not affect the measurement significantly when such a reagent is used.

It is clearly essential that if constancy of γ_X is to be achieved, the concentration of the inert electrolyte, added to fix the ionic strength, will be large with respect to the concentration of determinand. Thus, samples and standards containing high concentrations of determinand (greater than about 0.2 M) cannot be satisfactorily treated to fix γ_X and it is best to restrict treatment of the samples and standards to such buffering and decomplexing as may be necessary. In these cases the calibration graph will be curved and more standards than usual will be necessary to define it; in addition, variations in the composition of the samples, other than changes in determinand concentration, will give rise to inaccurate interpretation of the graph due to changes in γ_X. Allowance can be made for such inaccuracies by plotting several graphs for the different background compositions and determining the exact composition before each analysis.

At the other end of the concentration scale, as the determinand concentration in standards falls below 10^{-4} M, their preparation by serial dilution techniques becomes progressively less satisfactory owing to the solutions becoming progressively less stable and more subject to contamination. For the calibration of pH electrodes, this problem is overcome by the use of buffers. No competent chemist would attempt to calibrate a pH electrode at pH 7 in hydrochloric acid diluted to 10^{-7} M but would use instead a standard phosphate buffer of that pH. Similarly, for the calibration of other ion-selective electrodes, anionic and cationic buffers may be used with advantage. Buffers for the control of hydrogen ions and other ions are the subject of a book by Perrin and Dempsey.[9] Such buffers have been widely used outside the field of ion-selective electrodes for some time, especially in biological studies. Working with the fluoride-selective electrode, Baumann[10] used fluoride ion buffers consisting of fluoride complexes of H^+, Zr^{4+}, Th^{4+} and La^{3+} to study the behaviour of the fluoride electrode in very low fluoride concentrations. She showed that whereas when solutions prepared by serial dilution are used the effective lower limit of detection is about 10^{-6} M, when fluoride ion buffers are used Nernstian response is obtained right down to $10^{-9.5}$ M. Blum and Fog[11] proposed the use of metal ion buffers, of the type previously used by Chaberek and Martell,[12] for the calibration of electrodes and in particular demonstrated the efficacy of buffer solutions containing various proportions of cupric ions to EDTA or nitrilo-triacetic acid (NTA) for calibrating copper electrodes. An ionic strength of about 0.1 was maintained which helped to keep constant the activity coefficient of the free cupric ions in equilibrium with the Cu(II)/EDTA or Cu(II)/NTA complex. The values of pCu for the different buffers were calculated from the published stability constants by Ringbom's method.[13] Hansen, Růžička and their co-workers have formulated buffers for Ca^{2+},[14] Cd^{2+},[15] and Pb^{2+}[16] and used these for calibrating electrodes; the precise formulae of these buffers are not given in most cases, although the principles of the methods used to prepare them are expounded. These buffers are pH-sensitive[9,12,17] and therefore contain pH buffers to stabilize the metal ion activity.

These buffers are not yet commonly used, but deserve more study; the reservations already mentioned should be borne in mind.

Most electrode evaluations described in the literature include calibration graphs produced from results with 'standard' solutions prepared by serial dilution, occasionally even below 10^{-6} M. The uncertainty in the results from the solutions of lowest concentration is frequently substantial and nearly always unspecified. The degree of instability of these solutions is dependent on several parameters, such as the surface area to volume ratio of the solution container, the solution composition, the susceptibility of the determinand to complexation or biological attack, and the material of the container; thus, it is often impossible, using the data in the literature, to make a realistic comparison between the performances of different electrodes at these concentrations. Quoted values of limits of Nernstian response or, particularly, limits of detection should be viewed with circumspection. However, the use of buffers offers the possibility of preparing reproducibly the low concentration standards and, hence, of making realistic comparisons of different electrodes.

Unfortunately, buffer systems are not available for all the ions which may be detected by electrodes. It is improbable that buffers will be developed for such ions as sodium, potassium and nitrate; the uncertainties in the more dilute standards for these ions therefore have to be minimized by careful work. Results with the sodium electrode, described in the next chapter, show what can be achieved in solutions of extreme dilution in the most favourable conditions aided by the unreactivity of the sodium ions.

SELECTIVITY OF ELECTRODES

No ion-selective electrode responds exclusively to the ion which it is designed to measure, although it is often more responsive to this primary ion than to others. If another, interfering, ion is present at a concentration which is large with respect to the primary ion, the electrode response will have contributions from both the primary and interfering ions. The degree of selectivity of the electrode for the primary ion, A, with respect to an interfering ion, B, is expressed by the selectivity coefficient, k_{AB}; this is defined by the general equation for the electrode potential:

$$E = E^\circ \pm \frac{2.303RT}{n_A F} \log_{10} \left(a_A + \sum_B k_{AB} a_B{}^{n_A/n_B} \right) \qquad (3.10)$$

where n_A and n_B are the charges on ions A and B and E° is the standard potential of the electrode. The sign in the equation is positive when A is a cation and negative when A is an anion. This equation is applicable to nearly all electrodes.

When an electrode is very selective for A in comparison with B, then k_{AB} will be much less than unity. Conversely, if, as occasionally happens, the electrode responds preferentially to B rather than A, k_{AB} will be greater than unity. For example, the selectivity coefficient of the Orion calcium electrode for barium ions, k_{CaBa}, is 0.01; thus, the electrode is 100 times more responsive to calcium ions than barium ions. However, $k_{ClBr} = 300$, which shows that a chloride electrode gives a bigger response to bromide ions than to chloride ions. For the Orion divalent cation electrode, $k_{CaMg} = 1$; thus, the electrode is equally responsive to calcium and magnesium ions. To generalize, the electrodes with membranes based on inorganic salts and glass membranes have more extreme values of k_{AB} (i.e. $\ll 1$ or $\gg 1$) than the electrodes with membranes based on the organic ion exchangers.

The value of k_{AB} is never constant for all activities of A and B, although it is sometimes constant for a given ratio of the activity of A to B.[18] The electrodes based on the organic ion exchangers show particularly large ranges in k_{AB} for a given A and B; the results of Moody et al. with calcium electrodes given in Table 3.3 show a typical spread of values.[19] These variations in k_{AB} are

TABLE 3.3
Calcium electrode (Orion Model 92-20) selectivity coefficients[19]

Interferent	Concn range of interferent/M	k_{AB} Method 1	Method 2
Mg^{2+}	4.3×10^{-5}–5×10^{-3}	0.055–0.025	0.036–0.016
Ba^{2+}	5.0×10^{-3}–5×10^{-2}	0.033–0.011	0.034–0.011
Zn^{2+}	5.0×10^{-4}–5×10^{-3}	0.081	0.077–0.055
K^+	1.0	6.6×10^{-5}	1×10^{-4}
Na^+	1.0	1.7×10^{-4}	1.1×10^{-4}

associated with the mechanism of the electrode response and with the changing environment of the ions in solution. These factors also cause k_{AB} to depend on the method used to determine it. Hence, it is essential, especially for the electrodes based on ion exchangers, to know the details of the method of determining any quoted values of k_{AB}.

Several methods have been described for the experimental determination of selectivity coefficients. The methods fall into two categories: (i) separate solution methods, (ii) mixed solution methods.

In the separate solution methods the potential of the electrode under investigation is measured first in solutions containing the primary ion A with no B present and then in solutions containing the interfering ion B with no A present. k_{AB} is calculated from the activities of A and B, in the different solutions, and the electrode potentials. These separate solution methods of measuring k_{AB}, although simple to perform, cannot be recommended, as they do not give reliable results. The potential-determining processes at the surface of the electrode membrane are often substantially different in the separate pure solutions of A and B from those in mixed solutions of A and B. Thus, the method is unsatisfactory because the conditions of measurement of k_{AB} do not reproduce sufficiently closely the electrode environment in samples, which typically contain both A and B. The values of k_{AB} obtained by these methods often differ substantially from those obtained by one of the mixed solution measurements (see, e.g. Ref. 19).

In many cases the response of an electrode in pure solutions of an interfering ion is very slow, liable to drift and, hence, unsatisfactory. Reinsfelder and Schultz[20] found that when the steady state potentials of an electrode, based on an organic ion exchanger with a liquid membrane, were very different in the two pure solutions of primary ion and interfering ion at roughly equal activities, then the time taken by the electrode to reach these potentials, after transference between these two solutions, became very long. Furthermore, large initial transitory potentials were produced. As an extreme example, when a nitrate electrode was transferred from a 10^{-2} M nitrate solution into a 10^{-2} M hexafluorophosphate solution, a transitory potential of 150 mV was produced; when the electrode was returned to the nitrate solution, it took four days to reproduce its original steady state potential.

Mixed solution methods for determining k_{AB}

Mixed solution methods are always preferred to the separate solution methods. They entail the measurement of the electrode potential in a range of solutions containing different activities of A and B. For simplicity in interpreting the results, it is usual to prepare solutions either with a constant a_B and varying a_A or with a constant a_A and varying a_B. Conventionally the first method is preferred as it usually corresponds more closely to the situation in samples.

The second method has been used particularly when H^+ is the interfering ion,

B. In this case curves are produced showing the electrode potential in solutions of constant a_A but varying pH; a typical set of curves produced for a lead electrode[21] is shown in Fig. 3.2. Although values of k_{AH} may be calculated for each curve, since k_{AH} is not a constant it is more usual to use the curves directly to determine the working range of an electrode in samples of different pH. It is, of course, important to choose for the experiments pH buffers which do not interfere with the electrode response. Wilson, Haikala and Kivalo[22] argue cogently that this second method is not as satisfactory as the more usual method of varying a_A at constant a_B, even when B is H^+. In most analyses the pH range of a particular set of samples is very limited and indeed often adjusted by the addition of a constant amount of pH buffer. Thus, straightforward calibration graphs at a range of fixed pH values provide the most useful and relevant information on the selectivity of the electrode with respect to hydrogen ions. Wilson *et al.* determined k_{NaH} for sodium glass electrodes at several pH values in a range of buffers prepared from salts of tris-(hydroxymethyl)methylamine (tris buffers): this buffer system was chosen because the constituents interfere so little with the electrode response.

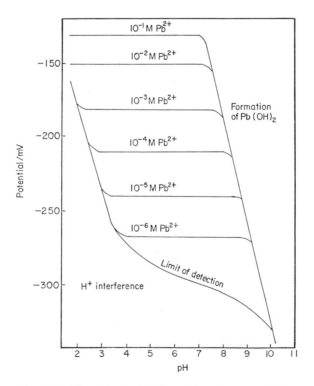

Fig. 3.2   The effect of pH on the response of the lead-selective electrode to a range of concentrations of $Pb(ClO_4)_2$ solutions[21]

If a range of solutions with constant a_B and varying a_A are prepared and the electrode potentials measured in these solutions and plotted against pA, a curve of the type shown in Fig. 3.3 is usually obtained. (See Chapter 5 for alternative behaviour.) The lower limit of a_A in these solutions should be set at the limit of Nernstian response of the electrode in a pure solution of A; otherwise the results become ambiguous. In the region PQ the electrode is responding in a

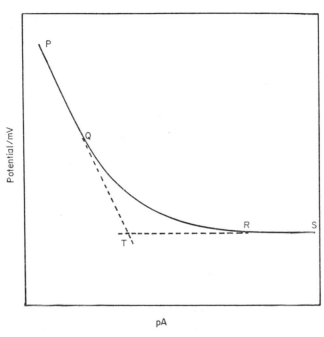

Fig. 3.3 Curve obtained for selectivity coefficient calculations
by varying the activity of the primary ion, A, in the presence
of a constant background of the interferent, B

Nernstian manner to the primary ion, A. As a_A decreases, the electrode potential is increasingly affected by the constant activity of B and in the region QR the electrode shows a mixed response to both A and B. From R to S the electrode is responding entirely to the constant a_B and the effect of the decreasing a_A is not detectable. Ideally the line RS will be straight and, as an example, is virtually straight when the electrode is a bromide-selective electrode, A is Br^- and B is I^-. However, as described above, with liquid ion-exchange electrodes, in pure solution of B the potentials may drift and show transitory peaks: in such cases RS will not be straight. The performance of most electrodes in the region RS lies between these two extremes.

There are several methods for calculating k_{AB} from the data presented in Fig. 3.3 and these will be described in turn.

Method 1. The first method depends on finding graphically the point T at which the electrode is responding equally to both ions, i.e. $a_A = k_{AB}a_B{}^{n_A/n_B}$ (from Eq. 3.10). If the line RS is straight and parallel to the abscissa, then T is the point of intersection of the extrapolations of PQ and SR as shown. k_{AB} may then be calculated from the activity of A at point T, $_Ta_A$, and the constant a_B by means of the equation

$$k_{AB} = \frac{_Ta_A{}^{n_B}}{a_B{}^{n_A}} \tag{3.11}$$

This method is only suitable if RS is a straight line.

Method 2. This more generally applicable method does not depend on the form of RS, but instead relies on PQ and QR. Again from Eq. (3.10), both ions are contributing equally to the electrode response when

$$a_A = k_{AB}a_B{}^{n_A/n_B} \tag{3.12}$$

If the activity of A at which this equality occurs is a'_A, and the activity of B is a'_B, then the electrode potential, E, is given by

$$E = E° + \frac{2.303RT}{n_AF} \log_{10}(a'_A + k_{AB}a'_B{}^{n_A/n_B})$$

$$= E° + \frac{2.303RT}{n_AF} \log_{10}(2a'_A)$$

The response of the electrode in the absence of B is given by the extrapolation of PQ as far as the limit of Nernstian response. The difference between the electrode potentials in solutions of A with activity a'_A with and without B at activity a'_B is therefore given by

$$\Delta E = \frac{2.303RT}{n_AF}(\log_{10} 2a'_A - \log_{10} a'_A)$$

$$= \frac{2.303RT}{n_AF} \log_{10} 2$$

$$\simeq \frac{18}{n_A} mV \quad \text{at } 25°C$$

Thus by finding on the graph the activity of A at which the experimental line QR differs from the extrapolation of PQ by $18/n_A$ mV (as in Fig. 3.4) the activity a'_A is determined. k_{AB} is then calculated by substitution into the equation

$$k_{AB} = \frac{a'_A{}^{n_B}}{a'_B{}^{n_A}} \tag{3.13}$$

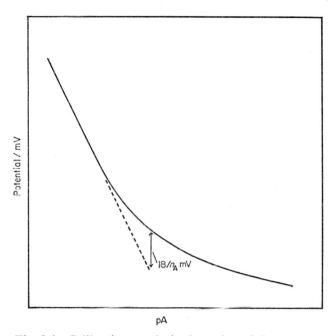

Fig. 3.4   Calibration graph, in the region of the response
limits, showing the method of determining the limit of
detection of the electrode

In a case such as the interference of SCN^- on the bromide electrode (see Fig. 5.4, curve b) this treatment is not possible and method 1 should be used.

Method 3. A least squares curve fitting procedure has been proposed by Wilson *et al.*[22] for fitting experimental data to Eq. (3.10) and determining both k_{AB} and $E°$. This method should, as these authors suggest, give the best fit of the data. They applied this method to the determination of k_{NaH} for various commercial sodium electrodes and found it to be satisfactory, except for some less selective electrodes which suffered such large interference that $E°$ changed during the experiment. The values of k_{NaH} determined in this way were 10–20% higher than those determined by method 2.

Method 4. Finally, the method of Srinivasan and Rechnitz[23] consists of fitting the results into a form of Eq. (3.10) algebraically rearranged to produce a linear graph. Unfortunately, this linearized form is rather complex and there is no evident advantage to compensate for the tedious calculation.

It will be apparent from the preceding discussion that seldom do any two methods yield the same value of k_{AB} from a given set of data, as exemplified in Table 3.3. This should not, however, be viewed as a limitation in the method of calculation but rather as an example of a limitation in the usefulness of the concept of k_{AB} as defined by Eq. (3.10). It may well be emphasized again that for any quoted value to be of use and comparable to other values, full details must be available of both the method of calculation and the experimental procedure used to derive the data (e.g. values of a_B, etc.).

Thus, values of k_{AB} must be treated with caution and are often not so useful in assessing electrode selectivity as the original curves of the type shown in Fig. 3.3. If a family of these curves is produced for different values of a_B, the analyst can see at a glance the likely level of interference expected, or the amount of interferent which can be tolerated in given samples. The use of values of k_{AB} is often limited in practice, owing to incompleteness of information, to the estimation of orders of magnitude of selectivity and to the ranking of k_{AB} values for different interfering ions, B. The selectivities of different electrodes may only be compared if the experiments are performed identically for each electrode.

In recent years much progress has been made in the development of theories explaining the selectivity behaviour of electrodes and allowing approximate values of k_{AB} to be predicted. Discussion of the selection mechanism of each type of electrode will be dealt with in the relevant chapters.

RESPONSE LIMITS

The detection limit of an ion-selective electrode has been defined in almost as many ways as there are electrodes. The lower limit of Nernstian response, or Nernstian limit, in contrast, is less ambiguous; it may be defined as the lower activity of the determinand, A, at which the Nernstian plot of electrode potential against $-\log_{10} a_A$ begins to depart from linearity. It may thus be determined directly from the calibration graph. This Nernstian limit is the more important criterion of electrode performance as it specifies the lower limit of the most useful range of the electrode. Below this limit, electrode response becomes progressively more and more irreproducible and regular use of the electrodes in this region, between the Nernstian limit and the limit of detection, is progressively less satisfactory; exceptions include the case of such very well-behaved electrodes as the silver chloride-based chloride electrode which may be used satisfactorily down to very close to the limit of detection.[24] Inevitably, as the electrode calibration graph becomes curved, many more calibration points are required to define the curve shape and the calibration will need to be repeated more frequently.

IUPAC provisionally recommend[25] that the detection limit be defined as the concentration of the determinand, A, at which the electrode potential deviates by $18/n_A$ mV from the extrapolation of the linear portion of the calibration graph. The definition is illustrated in Fig. 3.4. The definition is reminiscent of method 2 for calculating values of k_{AB} and emphasizes the fact that detection limits are often set by the concentration of interfering ions which cannot be eliminated from the sample background. Satisfactory though this definition appears in general, examples of its use are difficult to detect. However, listing other definitions does not seem worthwhile, since the detection limit, however defined, is of restricted practical value except as a very approximate guide.

The detection limit and the Nernstian limit depend on both the properties of the electrode and the conditions of measurement. As examples of the first factor, the response limits of calcium electrodes and chloride electrodes are set by the finite solubilities of the membrane material in the sample. When the electrodes are immersed in a sample, the membrane material dissolves into the sample until the equilibrium indicated by the solubility product is reached: the equilibrium activity of the determinand thus generated into a pure solution of background electrolyte without added determinand determines the limit of detection. For example, the solubility product of silver chloride is approximately $10^{-10}\,mol^2\,l^{-2}$ at 25 °C; hence, the equilibrium activity of chloride ions in a sample containing no interferents, silver ions or chloride ions is approximately $10^{-5}\,mol\,l^{-1}$. This value is thus the detection limit of a silver chloride-based chloride electrode and the limit is well defined.

The detection limit may also be set by the activity of defect ions in the membrane material: for example, the activity of defect silver ions in silver sulfide limits the sensitivity of the silver/sulfide electrode. Another electrode property which in some cases effectively limits the electrode response is the response time; this will be discussed later in the chapter.

More usually the response limits of an electrode depend, like values of k_{AB}, very much on the conditions of measurement. They will clearly be affected by the efficiency with which the available water and reagents can be freed from traces of determinand and interferents. In the case of the sodium electrode, as discussed in the next chapter, the practically determined detection limit appears to be largely determined by the purity of the available water and interference from the buffers used to prevent hydrogen ion interference; in addition, the instability of the very dilute concentrations of sodium is another important factor influencing the apparent limits. As already mentioned, the apparent limit of detection of electrodes when used in anion or cation buffers is often many orders of magnitude lower than when the electrodes are used in serially diluted solutions; hence, in these cases the measured limits of detection again depend on the conditions of measurement.

Other examples of these effects may be seen with gas-sensing probes. The measured response limit of an ammonia probe is set by the purity of the available water; it is very difficult to reduce the concentration of ammonia in water below $0.01\,mg\,l^{-1}$ (6×10^{-7} M) and, hence, it is difficult to test the response of the probe below this level. The nitrogen oxide probe when used under ambient conditions has a detection limit set by the interference from carbon dioxide in the water. In such cases the limit of detection becomes equivalent to a'_A (Eq. 3.13). The detection limit is apparently much lower if the experiments are carried out under nitrogen.

Response limits are also dependent on temperature and solvent. To take the example of the chloride electrode again, the detection limit is set by the solubility product of the silver chloride; hence, the detection limit changes from 1.8×10^{-5} mol l^{-1} at 25 °C to 5×10^{-6} mol l^{-1} at 5 °C. If, furthermore, the sample solvent is changed from water to acetic acid, a solvent in which silver chloride is less soluble, the detection limit drops even more.

To summarize, the response limits of an electrode or probe are always dependent on the conditions in which they are measured, and these conditions must be described if the value is to be of use. The limits can often be substantially altered by such means as purification of water and reagents or variation of temperature or solvent. Furthermore, the response limits are very dependent on membrane composition; hence, different electrodes for the same determinand may have substantially different limits of detection.

ASSESSMENT OF ELECTRODES AND PROBES

There are many points against which the performance and utility of a particular electrode or probe may be assessed for comparative purposes or to judge its usefulness in a particular application. The most important are as follows:

   (i) Response range and slope
  (ii) Selectivity
 (iii) Stability and reproducibility
 (iv) Response time
  (v) Sensitivity to temperature, pressure, light, etc.
 (vi) Frequency and ease of maintenance
(vii) Mechanical design
(viii) Availability
 (ix) Cost and lifetime

The points are not listed in order of importance since many are interdependent: as two examples, poor stability may be due to poor design and the response time is dependent on temperature. Each of the above points will be commented upon in turn.

(i) The response range of ion-selective electrodes and gas-sensing probes is very large compared with that of most analytical methods. Nearly all electrodes respond over at least four decades of concentration in a virtually theoretical manner and some, such as the sodium electrode, respond over eight decades of concentration. Nevertheless there are applications for which electrodes and probes are not sufficiently sensitive, such as the measurement of cyanide in treated effluents. As mentioned in the previous section, it is best to choose an electrode and so to fix the experimental conditions that measurements are made at concentrations higher than the Nernstian limit of the response.

Frequently more than one type of electrode is available to measure the same ion. These may include electrodes working either on the same principle but with different active membrane materials (e.g. the different copper electrodes with Ag_2S/Cu_xS membranes and $CuSe$ membranes) or on different principles (e.g.

the three ammonium sensors, the glass electrode, the monactin/nonactin electrode and the ammonia probe). Then each type must be assessed against the nine points listed above for suitability in a particular application, but the primary consideration will be whether the response range of a chosen sensor covers the range of concentrations of the sample.

Many of the electrodes based on inorganic compounds and glasses work satisfactorily in concentrations from the Nernstian limit right up to saturated solutions, although in the very strong solutions interpretation of the potentials becomes more difficult because of problems with liquid junction potentials and complexation on the electrode surface. Also, if concentrations are to be measured in strong solutions, uncertainty in the activity coefficient is another source of difficulty. The electrodes based on organic ion exchangers or neutral carriers, although they will respond in strong solutions, show increasing instability and concentration hysteresis in solutions of determinand more concentrated than 10^{-1} M. Similarly the gas probes become less satisfactory above 10^{-2} M and become increasingly non-Nernstian: they also show increasing concentration hysteresis as the concentration of the determinand approaches that of the internal filling solution. When the concentration of the determinand in a sample is too large for direct measurement, the sample may be diluted before measurement, although this is not possible in those rare cases in which dilution shifts some equilibrium of interest in which the determinand is involved.

(ii) The selectivity of electrodes and probes has been considered in general terms earlier in this chapter. The selectivity of these devices is, of course, central to their design and is one of the most important criteria in their assessment for a particular application. Before purchasing or using any electrode, the user should always be aware of the species which interfere with the electrode response; electrodes offered for sale without either these data or sufficient information on the electrode formulation to enable the data to be deduced, should be viewed with extreme caution by those analysts who wish to analyse samples and who are not interested in ion-selective electrode technology for its own sake.

In addition to the reversible interference represented by Eq. (3.10), some interfering species react irreversibly with the electrode membrane and poison it. As an example, mercuric ions react with the membrane of the silver sulfide-based electrodes (e.g. the silver/sulfide and copper selective electrodes) and form a coating of mercuric sulfide on the surface; the response is then destroyed and can be restored only by scraping off the mercuric sulfide.

High concentrations of interferents present problems, especially with the electrodes based on organic ion exchangers. Even if the electrodes are immersed in the interferent for a short time, the interfering ions migrate into the membrane and cause a shift in E°. If the sample is then changed to one not containing the interfering ions, these ions come out of the membrane only slowly and short-term hysteresis is observed. This behaviour is the basis of one of the arguments, previously discussed, against using separate solution methods for determining k_{AB}, especially in the case of the electrodes based on organic ion exchangers.

Gas-sensing probes are much more selective, in general, than ion-selective electrodes, as they suffer chemical interference only from other gases in solution. Some hysteresis effects may be observed after prolonged immersion of these probes in interferents.

(iii) The stability and reproducibility of electrodes and probes are primarily controlled by their environment, in particular by temperature; they always tend to be most stable and reproducible in solutions of high ionic strength containing large activities of determinand. But other important factors are the susceptibility of the sensors to hysteresis after large changes in concentration, after exposure to very high concentrations of determinand or interferents and after changes in temperature. As usual, the electrodes based on organic ion exchangers, especially those with liquid membranes, are most prone to these problems.

Instability may also be caused by excessive susceptibility of the electrode to static electricity; this may result from poor electrode design (wrong choice of plastics, etc.), faulty connexions inside the electrode or poor-quality cable. It is all too common to see one or two types of ion-selective electrode swathed in earthed aluminium foil in order to obtain sensible results, and one paper[26] even recommends boiling water in the laboratory to raise the humidity and so reduce noise. A further cause of instability, which is not the fault of the electrode, results from the small number of types of plug found on new electrodes and the large number of types of pH meter socket; these, coupled with the apparently frequent inability of the average analyst to fix a coaxial plug correctly on to a cable, have resulted in too harsh a verdict on many an electrode.

Because of the relatively poor accuracy and precision of electrode methods, it is particularly important to design experiments so that stability and reproducibility are optimized and to use the most satisfactory type of electrode. A good electrode, used under the right conditions, should show negligible drift of E° between successive calibrations (with calibration ideally once a day) and should be able to reproduce potentials to ± 0.1 mV. Some data on the stability and reproducibility of specific electrodes are given in the relevant chapters.

A substantial proportion of all problems apparently due to instability and irreproducibility of an electrode may be traced instead to the reference electrode and, in particular, to the liquid junction.

(iv) The definition of the response time of an electrode as provisionally given by IUPAC,[25] was stated in Chapter 1. Unfortunately, very few people have ever used this particular definition, and each group of workers has defined the response time in their own way. It is to be hoped that this very reasonable proposal by IUPAC will receive support. Values of response time quoted in the literature do not usually permit any more than a rough comparison of different electrodes, since the bases of the calculations are nearly all different, and often insufficient practical information is given on the circumstances of the measurement, upon which so much depends.

The measured response times of all types of electrode vary to some extent with the linear flow rate of the sample across the membrane; thus, for solid state halide electrodes, the response time may vary from a few hundred milliseconds in an experimental situation where there is a high linear flow rate to several tens of seconds in a solution of the same concentration slowly stirred in a beaker. The response time after a concentration change also depends on the experimental method by which that change is produced. If, for example, a step concentration change is produced in a sample by the rapid addition with fast mixing of a strong aliquot of determinand, the response will be much faster than after a transfer of the electrode between two beakers of solution of the same initial and final concentrations. In the second case the cell is made open-circuit, the equilibria on the membrane surface are upset and the liquid junction of the reference electrode is destroyed: the time taken for equilibrium to be regained in the whole system (cell + measuring system) is much greater, often at least ten times greater, than if the cell had not been disrupted.

The response times of most electrodes are sufficiently short for most analytical purposes. Typically electrodes take just one or two minutes to reach equilibrium; thus, the measurement time compares very favourably with most other analytical techniques.

The acceptability of the time taken for an electrode to respond depends on the particular application. For example, the response time of a sodium electrode in a 10^{-7} M sample of sodium ions in a power station analyser is about thirty minutes, but this is quite acceptable since minute-by-minute control is not required, only monitoring of the steady build-up of sodium ions. As another example, the response time of a sulfur dioxide probe in the analysis of beer containing 20 mg l^{-1} (3×10^{-4} M) sulfur dioxide is about five minutes, which is fast compared with the hour needed for analysis by a conventional distillation method. However, in manual procedures response times in excess of ten minutes make the analyses excessively tedious, and when an electrode is used in an automatic laboratory analyser, a response time of less than a minute is often required in order not to limit the rate of operation of the analyser.

Response times of sodium electrodes are particularly dependent on temperature, as shown by Wilson et al.;[27] the response times of other electrodes may also be expected to be temperature dependent but published results on this are few.

Factors affecting the response time have been discussed by Ryan and Fleet;[28] they conclude that three main factors determine the magnitude of the response time: (1) The type of membrane. The response speeds of the different types are generally in the order: solid state membranes > PVC membranes > liquid membranes (the last two change places for membranes based on neutral carriers, as described in Chapter 6). (2) The rate of change of solution activity. Rapid solution movement on the electrode surface helps to reduce response times. (3) The presence of interferents. Interferents generally slow the response.

(v) The effects on electrodes and probes of changes in temperature, pressure and incident light need consideration. Fortunately, the effects of changes in pressure or light intensity are usually negligible and are only significant with one or two electrodes. Changes in atmospheric pressure affect the gas-sensing probes by altering the Henry's law constant. In addition, pressure differences across the membrane, when the inside of the probe is at atmospheric pressure and the membrane is immersed in a sample in a sealed system, can alter the electrolyte film thickness and affect the response (see Chapter 7). Light has a small effect on some electrodes based on silver salts, but unless these electrodes are exposed to strong ultraviolet light, the effect may usually be ignored. The admixture of silver sulfide with silver chloride has been reported[29] to reduce the sensitivity of chloride electrodes to light.

The temperature coefficient ($dE°/dT$) of most electrodes is appreciable and thus knowledge of its magnitude is important so that the degree of temperature control required to maintain adequate electrode stability may be calculated. In some cases it is possible to match the temperature coefficient of the ion-selective electrode to that of the reference electrode so that the cell is approximately temperature-insensitive in respect to shift in $E°$. Correction must still, however, be made for the change in response slope of the ion-selective electrode with temperature. The isopotential concept may also be used to determine ways of minimizing temperature effects. Negus and Light[30] have shown that changing the internal filling solutions of electrodes or changing from isothermal to non-isothermal cells can reduce temperature effects: the authors list the effects of temperature on a number of cells.

(vi) The frequency at which an electrode or probe requires maintenance and the time taken to complete the maintenance are important, since their product gives the time the electrode is out of service. This is particularly important for electrodes used in continuous analysers. In the early days of ion-selective electrodes some of them, particularly the liquid membrane electrodes, were troublesome to service. However, in the last two or three years an increasing number of commercial electrodes of this type have interchangeable tips which may be changed in a very short time and which avoid the necessity of handling the obnoxious ion exchangers. Most of the electrodes based on inorganic salts require no maintenance other than an occasional clean to remove accumulated debris or coatings from the membrane surface. Glass electrodes require soaking every two or three months in a 0.1 M solution of the determinand for a few hours, especially if used in very dilute solutions. The electrodes based on organic ion exchangers or neutral carriers require servicing about monthly on average, although some of those with the ion exchanger embedded in a plastics have shorter lifetimes as they contain so little of the ion exchanger. Gas-sensing probes also require servicing approximately once a month, although, like all the electrodes, the frequency depends on the conditions of use and storage. The frequency of maintenance is related to the electrode lifetime, as discussed in section (ix).

(vii) Good mechanical design plays an important role in ensuring that an electrode or probe is stable and reproducible, has a long life and is easily serviced. The features of the design which optimize these points are, of course, independent of the styling of the electrode or the commercial packaging, although no self-respecting manufacturer ignores these latter features. A very approximate non-chemical correlation does appear to exist between the care put into the chemical and mechanical design of a commercial electrode and the care put into the design of the packaging and the instruction manual. Electrodes should be constructed of chemically resistant and stable materials and be sufficiently robust to withstand prolonged use. The same considerations apply to the cables and plugs.

(viii) Although many electrodes and probes are readily available commercially, many apparently satisfactory ones reported in the literature are not. This is usually either because the market for these has been deemed to be too small to make commercial production viable or because further private investigation has suggested that the original authors were too optimistic in their report of performance or failed to report some disqualifying feature. Short lifetimes and narrow response ranges are common faults of such electrodes. However, on occasions the analyst may wish to make his own electrodes for particular applications. Guidance on electrode manufacture is given in Chapter 10.

The wide range and ready availability of commercial electrodes and probes, together with the simplicity of the equipment needed in conjunction with them, have been important factors in the rapid increase in the number of applications of the sensors. Many people not usually in the forefront of analytical development have been able to develop their own analytical methods based on the sensors. This contrasts strongly with the growth of other analytical techniques, such as atomic absorption spectroscopy, where the expense of the equipment precluded most work on applications until the level of confidence in the new technique was much greater.

(ix) The cost and lifetime of electrodes and probes are not related chemically but are nevertheless firmly linked in the analyst's mind. An electrode must last long enough to earn its keep. Its use will usually result in time saving through shortening the time taken for analysis and thus allowing more analyses to be performed in a given time; moreover, if electrodes are used, the analyses may be done by relatively unskilled personnel. Electrodes and probes also offer the possibility of continuously analysing samples which may be difficult to analyse by any other technique. Consideration of these points and the average electrode lifetime allows a calculation of the cost effectiveness of the devices to be made.

The electrodes based on inorganic salts usually last for two or three years, unless used in a specially aggressive environment, and are simply thrown away when they cease to function properly. Electrode failure usually occurs because the seal holding the membrane to the body begins to leak. Occasionally the electrodes fail because of membrane pitting (sometimes cured by polishing) or, if the electrode is maltreated, because of membrane breakage.

The glass electrodes also usually last for two or three years. If the slope of the calibration curve drops, it may usually be restored by soaking as described previously. Failure usually occurs as a result of membrane breakage or loss of insulation across the cable connexions in the top cap.

The ion-exchange electrodes fail on average after about one month's use. As this is an unacceptably short lifetime, they are designed so that the parts are replaceable: the running cost must therefore be assessed in terms of the price of the refills of ion exchanger. Usually electrodes are supplied with several refills and it is worth while ensuring that these are stored under the optimum conditions since they do not keep indefinitely.

The gas-sensing membrane probes which are commercially available are also supplied with spare parts, in this case spare membranes and internal filling solution, usually sufficient to give at least a year's service.

The lifetime of each electrode or probe will depend on how carefully it is used and stored. The electrodes based on inorganic salts are best stored dry, whereas the glass electrodes are best stored in a dilute solution of their primary ion. Gas-sensing membrane probes should be stored as described in the instruction manual, which usually involves standing them in solutions of the same osmotic pressure as the internal filling solution but free of determinand. The lives of the electrodes based on organic ion exchangers or neutral carriers are more dependent on usage than the other types. The average lifetime of one month may be greatly exceeded if the electrode is used only occasionally and is stored in a dilute solution of the primary ion. However, for those types of electrode which have only a very small amount of ion exchanger, such as those containing the ion exchanger in a thin plastics matrix, it is the author's experience that the lifetime of these in continuous flow analysis is reduced to just a few days per refill. The older type of Orion liquid membrane electrode lasts for at least a month in continuous flow analysers.

Consideration of all these points should allow an informed decision to be made on the suitability of ion-selective electrode or gas-sensing probe techniques in a given application. Most reputable manufacturers will be able to provide information on all these points about their products; careful following of the advice in their instruction manuals should allow the electrodes and probes to be used most effectively.

REFERENCES

1. R. G. Bates, *Pure Appl. Chem.* **36**, 407 (1973).
2. R. G. Bates, B. R. Staples and R. A. Robinson, *Anal. Chem.* **42**, 867 (1970).
3. R. H. Stokes and R. A. Robinson, *J. Am. Chem. Soc.* **70**, 1870 (1948).
4. R. A. Robinson and R. H. Stokes, *Electrolyte Solutions*, 2nd edn, revised, Chapter 3, Butterworths, London (1965).
5. R. A. Robinson, W. C. Duer and R. G. Bates, *Anal. Chem.* **43**, 1862 (1971).
6. J. Bagg and G. A. Rechnitz, *Anal. Chem.* **45**, 271 (1973).
7. A. K. Covington and R. A. Robinson, *Anal. Chim. Acta* **78**, 219 (1975).
8. P. L. Bailey and E. Pungor, *Anal. Chim. Acta* **64**, 423 (1973).
9. D. D. Perrin and B. Dempsey, *Buffers for pH and Metal Ion Control*, Chapman and Hall, London (1974).

10. E. Baumann, *Anal. Chim. Acta* **54**, 189 (1971).
11. R. Blum and H. M. Fog, *J. Electroanal. Chem.* **34**, 485 (1972).
12. S. Chaberek and A. E. Martell, *Organic Sequestering Agents*, Wiley, New York (1959).
13. A. Ringbom, *Complexation in Analytical Chemistry*, Interscience, New York (1963).
14. J. Růžička, E. H. Hansen and J. C. Tjell, *Anal. Chim. Acta* **67**, 155 (1973).
15. J. Růžička and E. H. Hansen, *Anal. Chim. Acta* **63**, 115 (1973).
16. E. H. Hansen and J. Růžička, *Anal. Chim. Acta* **72**, 365 (1974).
17. H. U. Wolf, *Experientia* **29**, 241 (1973).
18. R. P. Buck, *Anal. Chim. Acta* **73**, 321 (1974).
19. G. J. Moody, R. B. Oke and J. D. R. Thomas, *Analyst* **95**, 910 (1970).
20. R. E. Reinsfelder and F. A. Schultz, *Anal. Chim. Acta* **65**, 425 (1973).
21. J. W. Ross, in *Ion-Selective Electrodes*, R. A. Durst, editor, Chapter 2. N.B.S. Spec. Publ. No. 314, Washington, D.C. (1969).
22. M. F. Wilson, E. Haikala and P. Kivalo, *Anal. Chim. Acta* **74**, 395 (1975).
23. K. Srinivasan and G. A. Rechnitz, *Anal. Chem.* **41**, 1203 (1969).
24. K. Torrance, *Analyst* **99**, 203 (1974).
25. *Recommendations for Nomenclature of Ion-Selective Electrodes*, Appendices on Provisional Nomenclature, Symbols, Units and Standards—Number 43, IUPAC Secretariat, Oxford (January 1975).
26. T. G. Sommerfeldt, R. A. Milne and G. C. Kozub, *Comm. Soil Sci. Plant Anal.* **2**, 415 (1971).
27. M. F. Wilson, E. Haikala and P. Kivalo, *Anal. Chim. Acta* **74**, 411 (1975).
28. T. H. Ryan and B. Fleet, *Proc. Anal. Div. Chem. Soc.* **12**, 53 (1975).
29. J. W. Ross, M. S. Frant and J. H. Riseman, U.S. Patent 3,563,974 (Feb. 16th, 1971).
30. L. E. Negus and T. S. Light, *Instrum. Technol.* 23 (December 1972).

CHAPTER 4

GLASS ELECTRODES

Glass electrodes designed to measure pH have been in common use for several decades. From early in the history of their development it was noticed that in strongly alkaline solutions interference with the pH response was experienced. In 1934 Lengyel and Blum[1] showed that the introduction of Al_2O_3 or B_2O_3 into a simple pH glass greatly increased the response of the glass to sodium ions to the extent that it responded to sodium ions according to the Nernst equation. Later the research teams of Eisenman and Nicolsky worked on the formulation of glasses with optimized selectivity for monovalent cations such as sodium or potassium ions. Nicolsky[2] first presented the form of the Nernst equation, already discussed in its general form in Chapter 3, which describes the response of a hydrogen ion-selective electrode to both hydrogen and sodium ions:

$$E = E^\circ + \frac{2.303RT}{F} \log_{10}\left(a_{H^+} + k_{HNa}a_{Na^+}\right)$$

This equation is often referred to as the Nicolsky equation.

Most of the early glasses with optimized sodium response were made from mixtures of sodium oxide, aluminium oxide and silicon dioxide. The formulation recommended by Eisenman[3] had the composition 11 mol% Na_2O, 18 mol% Al_2O_3 and 71 mol% SiO_2; this is known by the abbreviation NAS 11-18. In recent years this and other soda-based glasses have been partly replaced by lithia-based glasses containing Li_2O, Al_2O_3 and SiO_2. For potassium-selective electrodes the glass NAS 27-4, also developed by Eisenman, may be used.

THEORY

The theory of the response mechanism of glass electrodes and work on glass compositions is thoroughly described and discussed in the book edited by

Eisenman.[4] The book by Mattock[5] has a useful chapter on the theory, properties and use of pH-sensing glass electrodes, most of which applies equally to other glass electrodes; moreover, the reprint volume *The Glass Electrode*[6] condenses a great deal of both practical and theoretical information. Buck has reviewed more recent work[7] and further developed the theory.[8] In view of this ample coverage, it is proposed to give only the briefest outline of the theory here, and the more enquiring reader is referred to the above-mentioned sources.

When a glass electrode is immersed in an aqueous solution, the glass membrane absorbs water and a swollen hydrated layer is formed on the surface. The thickness of this hydrated layer varies considerably, according to the composition of the glass and the nature (i.e. pH, etc.) of the solution in which it is immersed, but generally is in the range $10-10^{-2}$ μm; the depth of hydration of lithia-based glasses is generally much less than that of the soda-based glasses. The internal surface of the glass membrane, in contact with the internal filling solution of the electrode (which is a pH buffer), will also be covered with a hydrated layer. Once formed, these hydrated layers should not be allowed to dry out.

At the interface of the hydrated layer with a sample solution the monovalent cations on the glass, sodium or lithium ions usually, are exchanged and the sites are involved in an ion-exchange equilibrium with the ion to be measured (determinand). The surface thus acts as a cation exchanger with the number of exchange sites determined by the structure of the hydrated glass. The potential at this boundary varies with the activity of determinand according to the Nernst equation. The determinand ions on the surface sites will tend to diffuse down through the hydrated layer towards the dry glass; simultaneously there will be a flux of cations from the dry glass to replace those dissolved away at the outer surface of the hydrated layer. Consequently the junction of the hydrated layer with the dry glass is very diffuse.

The electrode potential is made up of a number of contributions apart from that at the phase boundary already described, but these other contributions are largely independent of the determinand activity. There is a time-dependent diffusion potential, due to the flux of monovalent cations from the dry glass, which appears as a very slow drift in the response and has little effect on the equilibrium potential of the electrode. There is also a potential across the hydrated layer due to the charge separation produced by the opposing fluxes of determinand ion and ions from the glass (cf. liquid junction potentials). Other contributions to the potential include the, ideally constant, diffusion and phase boundary potentials on the inner surface of the glass. Contributions from diffusion within the dry glass may be ignored. This diffusion probably takes place by means of a defect or jump mechanism in which interstitial cations play a major role; a detailed discussion is given by Doremus.[9]

A final contribution to the potential, which is independent of determinand activity but dependent on time, is the asymmetry potential. This may be defined ideally as the residual potential between the glass surfaces when the two solutions on either side of the membrane are identical but in practice is also used to encompass the effect of other factors which gradually change with time. The major cause of asymmetry potentials is thought to be differences in stress in the two glass surfaces as produced, for example, in a thin glass bulb when rapidly cooled. Appreciable asymmetry potentials may also arise if the two surfaces age at different rates, perhaps owing to exposure of the surface in contact with the sample to extremes of pH, or if the chemical compositions of the surfaces differ slightly as a result of the method of manufacture (as produced, for example, by the volatilization of some sodium from the outer surface of a bulb formed in a flame): in both cases the depth of the hydrated layer on the two surfaces will differ. With most modern glasses asymmetry potentials are small and, more importantly, relatively constant, so that drifts in $E°$ due to changes in these potentials are negligible unless the electrode is maltreated.

Thus, the electrode potential is made up of a number of constant terms and one variable term which gives rise to the Nernstian response. The response of the electrode in solutions containing several different ions, which are able to compete for the ion-exchange sites, is more complicated. Eisenman[10] has developed a theory which has been successful in explaining the selectivity of a glass in terms of the ratio of the mobilities of the competing ions within the glass, and also the free energy changes, which occur during the ion-exchange processes on the surface of the hydrated layer. The free energy change is considered to be composed of two terms, one describing the difference in hydration energies of the two competing ions and the other the difference in coulombic attraction between each ion and the ion-exchange site. The equation for the electrode response in the presence of interfering ions which results from these considerations is somewhat more complicated than the Nicolsky equation; this accounts for the fact that the values of k_{AB} for glass electrodes determined by the methods described in Chapter 3 are not constant. However, Buck[11] has shown that plots of k_{AB} against $\log (a_B/a_A)$ are often linear and give valuable information on the mechanism of response of particular glasses.

Glass electrodes which respond to sodium, potassium and other monovalent cations are to be discussed in this chapter. The glass electrodes for the measurement of Fe(III) and Cu(II), which have recently been developed, are discussed in Chapter 8.

PERFORMANCE OF SODIUM ELECTRODES

There is a large range of sodium-selective electrodes commercially available and nearly all use different formulations for the membrane glass. More than for any other ion-selective electrode, the choice of the best electrode for a particular application is difficult because the performances of all these electrodes are different. Fortunately, several detailed studies, some of them comparative studies, have been made of the electrodes. The most recent and comprehensive in the literature is that by Wilson, Haikala and Kivalo,[12] who compared electrodes manufactured by Beckman, E.I.L., Orion and Radiometer. There appears to be no published independent study of some commercial electrodes.

The most important characteristics of sodium electrodes will be described in turn.

Response range and slope

The measured response range of a sodium electrode is more dependent than that of most other electrodes on the conditions of measurement. For measurements in beakers the Nernstian range of all electrodes extends down to between 10^{-4} and 10^{-5} M; departure from linearity above this concentration would indicate interference due to contamination of samples by the reference electrode bridge solution or to insufficient conditioning of the membrane.

It has been shown that two lithia-based glass electrodes, the Leeds and Northrup 117,201[13] and the E.I.L. 1048 400[14] (formerly called GEA 33) will respond according to the Nernst equation down to 5×10^{-8} M ($1 \mu g l^{-1}$) and probably down to 5×10^{-9} M ($0.1 \mu g l^{-1}$) when used in a flow system with pH adjustment by means of ammonia gas or volatile amines, such as diisopropylamine, absorbed into the sample stream. The response limits of these glasses therefore appear to be limited not by the membrane glass in the measurable range but by the purity of available water (it is very difficult[14] to prepare water containing less than 5×10^{-8} M Na^+), by the response time and by interference from H^+ and whatever buffers (NH_3 or volatile amines) are added to raise the pH.

The slope S of the calibration graph approaches the theoretical value (59.2 mV per decade at 25 °C) more closely as the electrode ages. A new electrode will often give about 55.5 mV per decade but after two or three months this slope will probably have increased to at least 58 mV per decade. Wilson et al.[12] present interesting data on the variation with ageing of several electrodes; some of their data are given in Table 4.1. Although a steady change may be seen to have occurred during the $3\frac{1}{2}$ months, all these variations are insignificant on a day-to-day basis. Commercial electrodes are often supplied with the membranes immersed in a sodium solution contained in a teat; in such cases the electrode will be at least partially aged before use.

TABLE 4.1
Variation of the slopes of sodium electrodes with ageing[12]

Electrode	0^a Slope/mV per decade	2.5^a Slope/mV per decade	3.5^a Slope/mV per decade
Beckman 39278	56.3	57.5	56.3
E.I.L. 1048 200	54.4	57.7	58.1
Orion 94-11	55.4	56.9	58.1
Radiometer G502 Na	55.8	56.9	59.1

The figures represent the mean values of the slopes of three electrodes of each type.
^{a} Months from new.

Reproducibility and stability

The short-term reproducibility of the response of a sodium electrode is approximately $\pm 1\%$ of concentration or activity (± 0.3 mV), in the range 10^{-1}–10^{-3} M Na^+ solutions, if care is taken. In general, $\pm 2\%$ of concentration (± 0.5 mV) is readily achievable. The reproducibility becomes poorer as the sodium ion concentration decreases; Webber and Wilson[14] obtained a standard deviation of $0.4 \, \mu g \, l^{-1}$ at the $2 \, \mu g \, l^{-1}$ (10^{-7} M) Na^+ level. These reproducibility figures are, of course, strongly dependent on the analytical method and the type of sample. The form of the electrode also has some effect. Friedman[15] has shown that if a tubular membrane, through which the samples flow, is used, sodium ion concentrations may be measured with an average error of only 0.20%.

The values of $E^{\circ\prime}$ of electrodes vary considerably both between the different types of electrode and even between identical electrodes of the same type. The difference between the potentials of different types of electrodes is largely due to the use of different internal filling solutions in the electrodes. $E^{\circ\prime}$ also varies with time and this is evident as a drift in the calibration points. Wilson et al.[12] have reported the data given in Table 4.2, in which three of each type of

TABLE 4.2
Variation of $E^{\circ\prime}$ of sodium electrodes with ageing[12]

Electrode	0^a $E^{\circ\prime}$/mV	2.5^a $E^{\circ\prime}$/mV	3.5^a $E^{\circ\prime}$/mV
Beckman 39278	146.2, 136.4, 150.7	122.2, 110.6, 110.9	82.4, 74.0, 79.5
E.I.L. 1048 200	131.3, 102.8, 135.9	95.1, 67.4, 93.4	91.5, 63.7, 99.3
Orion 94-11	122.0, 123.6, 121.3	121.0, 123.3, 121.5	127.3, 128.8, 130.6
Radiometer G502 Na	217.0, 202.4, 191.2	168.9, 162.8, 148.8	162.8, 143.1, 126.0

^{a} Months from new.

electrode were tested. The Orion electrodes show much less variation than the other electrodes, possibly because of the form of the membrane, which is a flat disc instead of the usual bulbous shape. Although some of the changes in Table 4.2 are large, the rate of drift of $E^{\circ\prime}$ of all electrodes is negligible, if the normal calibration frequency and the reproducibility of measurements are borne in mind. In a review article Eisenman[3] illustrates the variation of response with time and the stability and reproducibility of an electrode made from NAS 11-18 glass.

Temperature effects

The temperature coefficients of the standard potentials of sodium electrodes have not been the subject of much study: they are primarily dependent on the internal filling solution of the electrode. Mattock[16] reports an approximate value for the E.I.L. electrode of between $+0.1$ and 0.2 mV $^\circ C^{-1}$ in the range

20–40 °C. Negus and Light[17] calculated values of $+0.63$ and $+0.87$ mV °C^{-1} for the isothermal and nonisothermal temperature coefficients from their results with a cell including the Foxboro sodium electrode and the silver/silver chloride electrode with a 1 M KCl filling solution.

Selectivity

The effect of each interferent will be considered in turn.

Hydrogen ions

The interference of hydrogen ions on the response of sodium electrodes has been widely investigated. The results provide ample evidence that the apparent selectivity constant is very dependent both on the experimental procedure and on the calculation used to determine it.

A summary of the results of the determination of k_{NaH} is given in Table 4.3 with the method used to calculate the data listed according to the classification in Chapter 3. The data of Wilson et al.[12] are especially useful as they permit direct comparison of electrodes and also use the most realistic measurement technique.

TABLE 4.3
Literature values of k_{NaH} for commercial sodium electrodes

Electrode	Method	Medium	k_{NaH}	Ref.
Beckman 39278	Separate solns	0.1 M NaCl/0.1 M HCl	29	18
Beckman 39137	Separate solns	0.1 M NaCl/0.1 M HCl	4.15	18
E.I.L. 1048 200	Separate solns	0.1 M NaCl/0.1 M HCl	1.83	18
Beckman 39278	Titration procedure		4×10^3	8
Beckman 39278	1	Tris-acetate buffer pH 4.60	121	12[a]
E.I.L. 1048 200	3	Tris-acetate buffer pH 4.60	8.6	12[a]
Orion 94-11	1	Tris-acetate buffer pH 4.60	45.3	12[a]
Radiometer G502 Na	3	Tris-acetate buffer pH 4.60	3.04	12[a]
Beckman 39278	3	Tris-maleate buffer pH 5.20	62.8	12[a]
E.I.L. 1048 200	3	Tris-maleate buffer pH 5.20	22.7	12[a]
Orion 94-11	3	Tris-maleate buffer pH 5.20	63.7	12[a]
Radiometer G502 Na	3	Tris-maleate buffer pH 5.20	10.1	12[a]
Beckman 39278	3	Bis-tris buffer pH 6.52	164	12[a]
Orion 94-11	3	Bis-tris buffer pH 6.52	237	12[a]

[a] Values quoted from Ref. 12 are the mean values from measurements with three electrodes of each type.

Interesting experiments with the Beckman 39278 electrode by Buck[11] have shown that k_{NaH} is only constant when a_{Na^+}/a_{H^+} is greater than approximately 10^7. When $a_{Na^+}/a_{H^+} < 10^4$, then k_{NaH} varies according to the equation

$$\log k_{NaH} = A + \log (a_{Na^+}/a_{H^+})$$

where A is a constant. This behaviour is predicted by the theory of electrode response developed by Buck et al.[8] The variation of k_{NaH} with temperature has also been determined; average values of k_{NaH} at the different temperatures are given in Table 4.4; much fuller details are given in the original paper. It is

TABLE 4.4
Variation of k_{NaH} with temperature for Beckman electrode No. 39278 (k_{NaH} values averaged from data in Ref. 8)

Temperature/°C	k_{NaH}
6	3.4×10^3
14	8.3×10^3
25	4.0×10^3
41	2.8×10^3

curious that these values of k_{NaH} for the sodium electrode are so much higher than the values found by other workers (see Table 4.3).

The effect of the hydrogen ion interference is to limit the pH range in which a sodium electrode may be used without error; how narrow this range is will depend partly on the value of k_{NaH} for the particular electrode. From their results, Wilson et al.[12] deduced the lower pH limits given in Table 4.5. As expected, when the sodium concentration decreases, the pH range free from interference also decreases. The upper pH limit is set by the buffer, which must be alkaline and sodium-free, as will be discussed in the section on applications.

TABLE 4.5
Lower pH limit for sodium electrodes[12]

	Electrode			
[NaCl]/M	Beckman 39278	E.I.L. 1048 200	Orion 94-11	Radiometer G502 Na
10^{-1}	5.5	4.5	4.5	4.0
10^{-2}	7.0	5.0	5.5	4.5
10^{-3}	8.0	6.0	7.0	5.5
10^{-4}	9.0	7.0	8.0	6.5

The effect of pH on the sodium responses of NAS 11-18 glass (used by Corning and Orion) and that on BH 104 glass (used by E.I.L.) have been compared by Mattock.[16] Also, the effect of pH on E.I.L. and Beckman electrodes (Types 39278 and 39137) have been compared by Phang and Steel.[18] These results were in accord with those of Wilson et al.[12] In general, all the results indicate that samples containing sodium ions must be buffered so that pH > pNa + 3 or 4.

Potassium ions

The published data on the interference of potassium ions on the response of sodium electrodes are less detailed than the data concerning hydrogen ion interference. The NAS 11-18 glass has been shown by Eisenman to have a value of $k_{NaK} < 10^{-3}$ at high pH but only $<3 \times 10^{-3}$ at neutral pH.[3] The time dependence of the selectivity of this glass was also studied, a separate solution method being used to determine the selectivity coefficient at pH 11.3. Starting with a new electrode the value of k_{NaK} was 8.3×10^{-4} after 200 s and decreased to 3.6×10^{-4} after 10^5 s (28 h).

Wilson et al.[12] have determined k_{NaK} for four types of commercial electrodes; the results are given in Table 4.6. They tested three electrodes of each type and some at two concentrations of potassium ions as shown. It can be seen that there are large variations in k_{NaK} between electrodes of the same type and also that the values of k_{NaK} determined at the higher of the two potassium concentrations were substantially lower than in the more dilute potassium solution. Eisenman[3] has shown how the value of k_{NaK} for a particular glass may be used as a general index of the properties of that glass in the presence of other cations such as H^+, Li^+, Rb^+, Cs^+, Tl^+, NH_4^+ and Ag^+.

Silver ions

All sodium electrodes are very sensitive to silver ions; indeed many are more sensitive to silver ions than to any other ions, including hydrogen and sodium ions. Eisenman[3] discusses and presents results on the silver response of sodium glasses such as NAS 11-18 and even recommends the use of this glass for measurement of low concentrations of silver ions, perhaps down to as low as 10^{-9} M. In practice now, such a measurement would usually be made with a silver/sulfide electrode. The response of a sodium electrode to a range of concentrations of silver, lithium and thallous ions has been investigated by Mattock and Uncles.[19]

The average values of k_{NaAg} for the four commercial electrodes studied by Wilson et al.[12] are given in Table 4.6.

Because of the high sensitivity of sodium electrodes to silver ions, the leakage of dissolved species through the liquid junction of a single junction silver/silver chloride reference electrode very seriously interferes with some determinations, especially of low sodium concentrations. Hence, it is preferable to use a calomel reference electrode with a sodium electrode.

Ammonium ions

Values of k_{NaNH_4} determined by Wilson et al.[12] are included in Table 4.6. The high concentration of ammonium chloride used was necessary because of the low level of interference.

TABLE 4.6
Selectivity constants of sodium electrodes (k values averaged from data in Ref. 12)

Electrode	Method	Medium	k_{NaK}	Method	Medium	k_{NaAg}	Method	Medium	k_{NaNH_4}
Beckman 39278	4	0.1 M KCl	2.0×10^{-2}	3	10^{-6} M AgNO$_3$	370	3	1 M NH$_4$Cl	1.5×10^{-4}
E.I.L. 1048 200	4	0.1 M KCl	9.7×10^{-4}	3	10^{-6} M AgNO$_3$	210	3	1 M NH$_4$Cl	5.1×10^{-5}
Orion 94-11	4	0.1 M KCl	9.2×10^{-3}	3	10^{-6} M AgNO$_3$	380	3	1 M NH$_4$Cl	6.1×10^{-5}
Radiometer G502 Na	4	0.1 M KCl	2.3×10^{-4}	3	10^{-6} M AgNO$_3$	110	3	1 M NH$_4$Cl	3.6×10^{-5}
E.I.L. 1048 200	4	0.5 M KCl	2.1×10^{-4}						
Radiometer G502 Na	4	0.5 M KCl	9.1×10^{-5}						

The figures represent mean values from measurements with three electrodes of each type.

The response of the Leeds and Northrup 117,201 electrode to ammonium ions has been studied by Eckfeldt and Proctor[13] in the context of low-level sodium determination (see Fig. 4.1).

Other species

Sodium electrodes respond to other monovalent cations such as Rb^+, Cs^+, Li^+ and Tl^+ and even, to a lesser extent, divalent cations such as Ca^{2+} and Mg^{2+}. Normally these ions do not interfere unless present in very large excess over the sodium ions. Buck[11] has reported computed values of k_{NaLi} for the Beckman sodium electrode.

The response of a sodium electrode in volatile amines such as di-isopropyl-amine has been studied;[13] the level of response is so low that the use of such amines in place of ammonia has been suggested for buffering low-level sodium samples. Webber and Wilson[14] studied the effect on the response of other substances likely to be present in condensates from power stations. Octadecyl-amine, a filming amine, produced a marked interference.

Response time

The response times of sodium electrodes have been measured in both strong and dilute solutions of sodium ions and at different temperatures. In no case has the definition of response time proposed by IUPAC been used; hence, the usage of each worker has to be considered.

Wilson et al.[12] measured the response times of four types of commercial electrode. Measurements were made at both 25 °C and 10 °C. They defined the response time as the time taken for the electrode potential to change by 95% of the difference in the equilibrium potentials of the electrode in the first and second solutions. The results are given in Table 4.7.

The response time of the E.I.L. sodium electrode in high-purity water has been measured by Diggens, Parker and Webber.[20] The electrode was mounted in a commercial continuous flow monitor. For sodium concentrations down to $2\ \mu g\,l^{-1}$ (10^{-7} M) they found that 90% response was achieved in less than

TABLE 4.7
Response times of sodium electrodes (results averaged from data in Ref. 12). Times for changes from 10^{-4} to 10^{-3} M Na$^+$. Solutions at pH 9.54 in ethanol ammonium buffer

Electrode	Response time/s	
	10 °C	25 °C
Beckman 39278	64	14
E.I.L. 1048 200	70	10
Orion 94-11	245	30
Radiometer G502 Na	287	49

These results represent mean values from measurements with three electrodes of each type.

five minutes; however for 100% response (the time to reach the final equilibrium potential), the response time is up to thirty minutes at the lowest concentrations and ten minutes at about 20 μg l^{-1} (10^{-6} M).

APPLICATIONS OF SODIUM ELECTRODES

The types of samples in which sodium electrodes have been used fit into three categories: biological fluids, high-purity water and other samples (mainly foodstuffs).

The determination of sodium in biological fluids has been extensively reviewed in chapters in several books.[15,21-23] Sodium micro-electrodes have been particularly important in widening the range of possible *in vivo* measurements; such measurements have been used for both research and diagnostic purposes. Good correlation has been demonstrated by several workers between the results of analyses using a sodium electrode and those using a flame photometer. Usually in the analysis of biological samples the samples are not buffered before measurement, and thus it is assumed that the pH is sufficiently high for interference by H$^+$ to be ignored; interference from other ions is not usually significant. Also, if no ionic strength adjuster is added (as when *in vivo* measurements are performed) and sodium ion concentrations are to be measured, allowance must be made for variations in the activity coefficient of the sodium ions. However, in such media it is often sodium ion activity that is required, in which case activity standards must be used for calibrating the electrode.

The determination of very low concentrations of sodium in high-purity water is a particularly important application of the sodium electrode. The analysis of these waters is essential in the control of the purity of boiler water and in monitoring the output of mixed-bed ion-exchange columns to detect breakthrough. The level of sodium in boiler feedwater in power stations may drop below 1 μg l^{-1} (5×10^{-8} M or pNa = 7.3); hence, to eliminate interference from H$^+$ the samples must be made alkaline (pH $\geqslant$ 11). The major problem in these applications is to achieve this pH without introducing sodium contamination from the buffer; hence, it is crucial to choose the right buffer system. It is essential to carry out all measurements in a flowing non-glass system to minimize the amount of sodium picked up from the environment: in such a system the reference electrode is placed downstream from the sodium electrode to eliminate interference from leakage of the bridge solution. Diggens *et al.*,[20] using the E.I.L. sodium electrode in a continuous-flow monitor, buffered the sample stream by sucking in ammonia vapour: they made satisfactory measurements down to 2 μg l^{-1} which correlated well with flame photometric analyses. It was envisaged that measurements at even lower concentrations would be possible. Webber and Wilson[14] also used the electrode at these levels; their results suggest a slight interference ($\leqslant$1 μg l^{-1}) due to electrode response to ammonium ions.

Eckfeldt and Proctor[13] experimented with various volatile amine buffers, which cause less interference than ammonia, for use in measurements at extreme dilution. They reported measurements with the Leeds and Northrup electrode, using di-isopropylamine as buffer, as low as $0.07 \, \mu g \, l^{-1}$ $(2 \times 10^{-9} \, M)$: as shown in Fig. 4.1, this amine interferes with the electrode response less than any

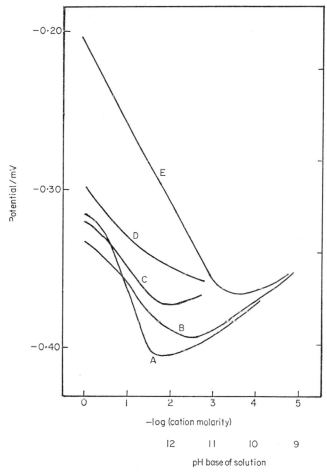

Fig. 4.1   The response of a sodium electrode to different base reagents at a range of concentrations, showing also the apparent sodium ion activity indicated by the response to the bases.[13] A, di-isopropylamine; B, dimethylamine; C, methylamine; D, ethylamine; E, ammonia

other tested. A notable achievement of these authors was the production of water containing less than $0.1 \, \mu g \, l^{-1} \, Na^+$. They remark specially that a pressurized reference electrode must be used to produce a stable liquid junction potential. In a later paper,[24] they demonstrated that the system is more sensitive if the flow cell is designed to produce a high sample velocity past the bulb of the sodium electrode. It is also advantageous[25] to keep the temperature as low as is consistent with a reasonable electrode response time as, because of the high temperature coefficient of the buffer, the reagent consumption to produce a given sample pH is lower at lower temperatures (see also Chapter 9).

Webber and Wilson,[14] in addition to comparing the results of analyses by an electrode method and by flame photometry, determined the precision and bias of a series of measurements. The standard deviations of their measurements ranged from $10 \ \mu g \, l^{-1}$ at a sample concentration of $230 \ \mu g \, l^{-1}$ to $0.35 \ \mu g \, l^{-1}$ at $1.8 \ \mu g \, l^{-1}$.

There are many applications[26] of the sodium electrode in samples containing much higher concentrations of sodium ions. Straightforward applications include the analysis of natural waters,[27] soil extracts[28] and beer. Sekerka and Lechner[29] have developed an automatic method for measuring sodium in the concentration range $0.1-100 \ \text{mg} \, l^{-1}$ in natural and waste waters. The standard deviation of results was typically about 5% and satisfactory agreement was obtained with a flame photometric method. The electrode methods have the advantage that the samples need not be clarified or filtered before measurement.

A more challenging application, which shows the power of electrode techniques, is in the method for the determination of salt in bacon developed by Halliday and Wood.[30] The authors pushed a sodium electrode and a reference electrode into macerated bacon samples and took a reading directly: no liquid was added. The results correlated satisfactorily with those obtained by the Volhard method. An even simpler method, which gave remarkably good results for most types of bacon, was to insert the two electrodes into holes cut in a roll of bacon with a cork borer. The sodium content and pH of the bacon were high enough for hydrogen ion interference to be negligible.

Sodium electrodes have also been used in fundamental work. Studies of the activity coefficients of sodium and chloride ions in sodium chloride solutions[31] and the activity coefficient of sodium ions in mixed sodium chloride and potassium chloride solutions[32] have been carried out by means of sodium electrodes. The electrodes have also facilitated investigation of ion association in solutions of sodium tetrametaphosphate and trimetaphosphate.[34]

A general-purpose buffer, which has been recommended by Mattock[16] for both buffering the samples to prevent hydrogen ion interference and fixing the ionic strength, is 0.5 M or 1 M triethanolamine adjusted to pH 7–8 by hydrochloric acid. It is essential to use a grade of triethanolamine which has a very low sodium content. The volume ratio chosen of sample to buffer will depend on the nature of the sample.

Sodium electrodes age slowly during use; as shown by Wilson et al.,[12] the value of E° shifts and the slope becomes more closely Nernstian. However, eventually the time comes (more rapidly if the electrode is continually used in low concentrations of sodium ions) when the slope begins to decrease, the response time lengthens and the response becomes more irreproducible. The response may usually then be restored to full efficiency by soaking the electrode for 24 hours in 0.1 M NaCl solution, followed by careful rinsing in distilled water. To maintain the electrode performance for as long as possible, it should be stored between readings in dilute sodium chloride solution of approximately the same concentration as in the samples to be analysed. In the case of electrodes for analysing high-purity water, after soaking in 0.1 M NaCl solution for 24 hours, the electrode should be soaked in frequently changed, recirculated, distilled and deionized water for several days. The lifetime of the electrode is inevitably limited by dissolution of the glass, which is accelerated if the application requires that the electrode is continually used in alkaline media.

PERFORMANCE OF AMMONIUM, POTASSIUM AND MONO-VALENT CATION ELECTRODES

The performances of these electrodes have been less thoroughly investigated than the performances of sodium electrodes. There are a few important applications of them but, in general, they are not widely used. This is largely because other sensors for the same determinands, e.g. the valinomycin-based potassium electrode and the ammonia probe, are more selective. However, these glass electrodes are usually much cheaper than the other sensors, and, in the case of the potassium electrode, have a lower detection limit than alternative sensors. Moreover, they are attractive in biological studies because of the relative inertness of the membrane.

In the absence of an appreciable amount of independent comparative data, Table 4.8 presents representative data drawn from manufacturers' information and studies of the performance of individual electrodes.[34-37]

TABLE 4.8
Performance of cationic glass electrodes (expressed as potassium electrode performance)

Response range:	typically Nernstian to about pK 6
pH range:	for no H^+ interference pH $>$ pK $+ 3$
Reponse order (typical):	$H^+ > Ag^+ > K^+ \sim NH_4^+ > Na^+ > Li^+$
Selectivity coefficients:	$k_{KNH4} \sim 1$, $k_{KNa} \sim 0.2$, $k_{KLi} \sim 0.06$, $k_{KRb} \sim 0.5$

APPLICATIONS OF AMMONIUM, POTASSIUM AND MONO-VALENT CATION ELECTRODES

A major application of the potassium electrode, like the sodium electrode, is in the analysis of biological fluids, a field which has been fully reviewed.[15,21-23] In most biological fluids there is an excess of sodium ions over potassium and a correction must be applied for the sodium interference; hence, the sodium ion concentration is often measured simultaneously by use of a sodium electrode.

A serious problem with the electrode is the poor selectivity to any one monovalent cation, and the difficulty of removing any which do not have to be measured. Hence, the most satisfactory applications are the analysis of those samples consisting of virtually pure determinand. Even so, pH adjustment is necessary, as with the sodium electrode; most usual buffers may not be used, but the triethanolamine buffer already mentioned for the sodium electrode is suitable.[34]

The most important application of such electrodes used as ammonium ion sensors is in the analysis of boiler feed waters with ammonium ion concentrations down to $10\,\mu g\,l^{-1}$ $(6 \times 10^{-7}\,M)$. The pH is adjusted by means of the triethanolamine buffer. An investigation of the method has been made by Goodfellow and Webber,[35] who studied the accuracy and also the interference caused by a range of species from lithium ions to n-octadecylamine. The results of analyses using the electrode method agreed to within 5% with the results from an indophenol blue method for samples with concentrations in the range 0.1–0.8 mg $NH_3\,l^{-1}$. Mertens et al.[37] have published a comparison of the two alternative ammonia sensors (the ammonia probe and the monactin/nonactin electrode): there is comparatively little to choose between the three sensors for this application.

Eagan and Dubois[38] compared an ammonium-selective glass electrode with an ammonia probe for analysis of ammonium ions in air-borne particles. Both worked satisfactorily but, in the experimental conditions used, the glass electrode worked down to lower concentrations.

Cation-selective electrodes have been used as sensors in enzyme electrodes; these systems will be discussed in Chapter 8.

The electrodes should be stored and reactivated in the same way as sodium electrodes but soaking them in a 0.1 M solution of the determinand. To minimize interference from the bridge solution of the reference electrode, it is advisable to use a mercury/mercurous sulfate reference electrode; alternatively, a double junction reference electrode may be used with a bridge solution either of similar composition to the sample or of 0.1 M HCl.[35]

REFERENCES

1. B. Lengyel and E. Blum, *Trans. Faraday Soc.* **30**, 461 (1934).
2. B. P. Nicolsky, *Acta Physicochim. URSS* **7**, 597 (1937).
3. G. Eisenman, *Advan. Anal. Chem. Instrum.* **4**, 213 (in Ref. 6).
4. G. Eisenman (ed.), *Glass Electrodes for Hydrogen and Other Cations*, Marcel Dekker, New York (1967).
5. G. Mattock, *pH Measurement and Titration*, Heywood, London (1961).
6. G. Eisenman, R. G. Bates, G. Mattock and S. M. Friedman, *The Glass Electrode*, Interscience, New York (1966).
7. R. P. Buck, *Anal. Chem.* **44**, 270R (1972); *Anal. Chem.* **46**, 28R (1974).
8. R. P. Buck, J. H. Boles, R. D. Porter and J. A. Margolis, *Anal. Chem.* **46**, 255 (1974).
9. R. H. Doremus, Chapter 4 in Ref. 4.
10. G. Eiseman, Chapter 7 in Ref. 4.
11. R. P. Buck, *Anal. Chim. Acta* **73**, 321 (1974).

12. M. F. Wilson, E. Haikala and P. Kivalo, *Anal. Chim. Acta* **74**, 395, 411 (1975).
13. E. L. Eckfeldt and W. E. Proctor, *Anal. Chem.* **43**, 332 (1971).
14. H. M. Webber and A. L. Wilson, *Analyst* **94**, 209 (1969).
15. S. M. Friedman, in *Methods of Biochemical Analysis*, Vol. 10. D. Glick, editor, p. 71 (in Ref. 6).
16. G. Mattock, *Analyst* **87**, 930 (1962).
17. L. E. Negus and T. S. Light, *Instrum. Technol.* 23 (1972).
18. S. Phang and B. J. Steel, *Anal. Chem.* **44**, 2230 (1972).
19. G. Mattock and R. Uncles, *Analyst* **87**, 977 (1962).
20. A. A. Diggens, K. Parker and H. M. Webber, *Analyst* **97**, 198 (1972).
21. E. W. Moore, Chapter 15 in Ref. 4.
22. R. N. Khuri, in *Ion-Selective Electrodes*. R. A. Durst, editor, Chapter 8. N.B.S. Spec. Publ. No. 314, Washington, D.C. (1969).
23. M. Lavallée, O. F. Schanne and N. C. Herbert (eds.), *Glass Microelectrodes*, Wiley, New York (1969).
24. E. L. Eckfeldt and W. E. Proctor, *Anal. Chem.* **47**, 2307 (1975).
25. E. L. Eckfeldt, *Anal. Chem.* **47**, 2309 (1975).
26. G. Mattock, in *Analytical Chemistry* 1962 (Proceedings Feigl Anniversary Symposium, Birmingham), p. 247. Elsevier, Amsterdam.
27. N. V. West and J. Pursey, reported in Ref. 26.
28. C. A. Bower, *Soil Sci. Soc. Am. Proc.* **23**, 29 (1959).
29. I. Sekerka and J. F. Lechner, *Anal. Lett.* **7**, 463 (1974).
30. J. H. Halliday and F. W. Wood, *Analyst* **91**, 802 (1966).
31. A. Shatkay and A. Lerman, *Anal. Chem.* **41**, 514 (1969).
32. R. Huston and J. N. Butler, *Anal. Chem.* **41**, 1695 (1969).
33. G. L. Gardner and G. H. Nancollas, *Anal. Chem.* **41**, 202 (1969).
34. G. Mattock and R. Uncles, *Analyst* **89**, 350 (1964).
35. G. I. Goodfellow and H. M. Webber, *Analyst* **97**, 95 (1972).
36. G. G. Guilbault, R. K. Smith and J. G. Montalvo, *Anal. Chem.* **41**, 600 (1969).
37. J. Mertens, P. Van den Winkel and D. L. Massart, *Bull. Soc. Chim. Belg.* **83**, 19 (1974).
38. M. L. Eagan and L. Dubois, *Anal. Chim. Acta* **73**, 157 (1974).

CHAPTER 5

ELECTRODES BASED ON
INORGANIC SALTS

Electrodes based on inorganic salts were the first of the new generation of ion-selective electrodes which succeeded the glass electrodes. In the mid-1960s Pungor and his colleagues produced silver and halide electrodes, the membranes of which consisted of silver salts, prepared by precipitation methods and dispersed in silicone rubber. These are reviewed in Ref. 1. The development of these electrodes was soon followed by the fluoride electrode of Frant and Ross[2,3] and also the silver and halide electrodes with membranes consisting of pressed pellets of the appropriate silver salt coprecipitated with silver sulfide. These in turn led to the development of the electrodes for the heavy metals copper, lead and cadmium. Thus, inorganic salt-based electrodes are now available for the determination of F^-, Cl^-, Br^-, I^-, S^{2-}, SCN^-, Cu^{2+}, Pb^{2+}, Cd^{2+} and Ag^+; the iodide electrode may also be used to determine CN^- and Hg^{2+}. A few other electrodes which fall into this category have been reported but are not commercially available.

These electrodes are constructed in various forms. The main points of variation between different electrodes for the same species are the construction of the membrane and the internal contact to the membrane.

Membranes are usually termed either homogeneous or heterogeneous. The homogeneous membranes consist entirely of the active material or materials pressed or machined into the required membrane shape, usually a disc; examples of this construction are the chloride-selective membrane consisting of a pressed pellet of coprecipitated silver chloride and silver sulfide and the fluoride selective membrane, which is a machined crystal of lanthanum fluoride. The heterogeneous membranes consist of an active material, sometimes but not always the same as in the corresponding homogeneous membrane, supported in an inert matrix such as silicone rubber, PVC or polythene. The amount of the supporting material used is kept to the minimum consistent with membrane strength in order to maximize contact between the particles of active material. Usually, but not always, the response mechanisms and performances of homogeneous and heterogeneous membranes prepared from the same active material are so similar that distinction between the different constructions is not important for the analyst. The variety of constructions which are found in the different commercial electrodes owes more to the patent laws and the predilections of individuals than to chemistry.

Contact must be made inside the electrode between the inner surface of the membrane and the cable, and made in such a way that the internal potentials produced are stable. With glass pH electrodes a chloridized silver wire is usually soldered to the cable and immersed in a chloride-containing pH buffer which fills the inside of the glass bulb. Thus, the glass contacts the buffer solution, and a stable potential is developed across the interface dependent on the buffer pH; also, the solution contacts the wire (internal reference electrode), and another stable interfacial potential is produced dependent on the chloride concentration in the buffer. An analogous method for contacting the membrane may also be used in the inorganic salt-based ion-selective electrodes; again it is customary to use a silver/silver chloride internal reference electrode but the internal solution is changed to one containing chloride and an ion to which the membrane responds, usually the primary ion for which the electrode is designed. The fluoride electrode, for example, is constructed in this manner. This has a silver/silver chloride reference electrode and an internal electrolyte consisting of 1 M potassium fluoride, saturated potassium chloride and saturated silver chloride.[2] In most of the early work with ion-selective electrodes the membranes were contacted in this way; however, more recently many have solid state internal contacts. For those membranes containing a silver salt, solid contacts are now used almost exclusively and these are formed by either embedding a silver wire into the inner surface of the membrane or sticking the internal contact on to this surface with an adhesive, such as epoxy cement, loaded with finely divided silver. Other materials which have been satisfactorily used for directly contacting membranes include mercury, carbon and platinum.

A final class of inorganic salt-based electrodes which may be used as sensors comprises the electrodes of the second kind, of which the commonest are those based on silver such as the silver/silver chloride electrode. These cannot, by current usage, be classed as ion-selective electrodes because the membrane of inorganic salt is usually porous and allows the sample solution to reach the base metal. Thus, the electrodes are sensitive to redox effects. However, they have been in use for many years, and in many applications where no redox systems are present are just as satisfactory as the corresponding ion-selective electrode. They may easily be prepared in the laboratory by forming a layer of the appropriate silver salt directly on to a silver wire or billet, by dipping into the molten salt or by electrolytic deposition (see Chapter 2).

The various types of electrodes are illustrated in Fig. 5.1.

Many different recipes for the preparation of the different electrode membranes are given in the literature. However, in practice only a few are used in commercial electrodes and even fewer of the resultant electrodes have been properly compared one with another. Orion Research Inc. have played a pioneering role in both the development and application of the majority of the inorganic salt-based electrodes and, in particular, have been in the forefront of the development of the solid state electrodes (electrodes with both solid membranes and solid internal contacts); as a result of all these factors they dominate the commercial electrode market in many parts of the world. Accordingly, the situation has arisen that when new varieties of solid state electrodes are evaluated,

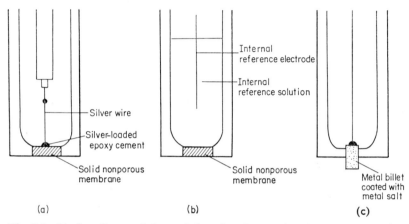

Fig. 5.1 Various forms of electrode based on inorganic salts: (a) ion-selective electrode with solid internal contact; (b) ion-selective electrode with internal solution and internal reference electrode; (c) electrode of the second kind

it is customary to compare their performance with that of the corresponding Orion electrode. This does not necessarily imply that the Orion electrodes are the best, merely that they are often a convenient reference for comparative purposes. The constitutions of the membranes of Orion electrodes and of some of the other satisfactory membranes are given in Table 5.1 (only the active materials are listed). In this chapter most of the performance characteristics given refer both to Orion electrodes and to others, unless otherwise stated, although most of the work reported has been done with Orion electrodes.

The theoretical principles governing the response mechanisms of different electrodes for the same primary ion are for the most part identical, although the membranes based on the mercury systems are an exception.

An electrode which has been reported but which is not included in Table 5.1 is the sulfate electrode developed by Mohan and Rechnitz:[4] the membrane is a mixture of $PbSO_4$, PbS, Ag_2S and Cu_2S; full details of the method of membrane preparation are given and the electrode is conditioned by soaking in lead nitrate solution. Although a sulfate electrode would be of great practical use,

TABLE 5.1
Active materials of electrode membranes

Primary ion	Orion electrode	Other homogeneous membranes	Heterogeneous membranes
F^-	LaF_3		—
Cl^-	$AgCl/Ag_2S$	Hg_2Cl_2/HgS, $AgCl$	$AgCl$
Br^-	$AgBr/Ag_2S$	Hg_2Br_2/HgS, $AgBr$	$AgBr$
I^-			
CN^-	AgI/Ag_2S	AgI	AgI
Hg^{2+}			
SCN^-	$AgSCN/Ag_2S$	$Hg(SCN)_2/HgS$, $AgSCN$	$AgSCN$
S^{2-}	Ag_2S	Ag_2S	Ag_2S
Cu^{2+}	Cu_xS/Ag_2S	$CuSe$	Cu_xS, Cu_xS/Ag_2S
Pb^{2+}	PbS/Ag_2S		PbS/Ag_2S
Cd^{2+}	CdS/Ag_2S		CdS/Ag_2S

this electrode is not available commercially and few reports of its application have yet been published; further details of its performance are awaited with interest.

THEORY

Theories describing the mechanism by which inorganic salt-based electrodes respond have been developed by several workers, and attempts have been made to show that the mechanism is analogous to that of glass electrodes. In general, it is assumed that when an electrode is immersed in a sample, interfacial potentials are developed which are associated with the ion-exchange processes on the membrane surface; in addition, diffusion potentials are set up due to the diffusion of the various ionic species within the membrane. Some authors have also postulated the formation of a hydrated layer on the surface of some membranes which is essential to their function; so far evidence for this is limited. In these fields experiments involving a.c. polarization of the membranes[5] or uptake of radioactive species into the membranes (see, e.g. Refs 6 and 7) have been informative.

In this section coverage is restricted to those theories developed to explain the more important performance characteristics of the electrodes, such as standard potentials, response ranges and selectivities. Details of the theories describing the mechanism of charge transfer through the membranes are, in general, not included. The fluoride electrode is dealt with separately at the end of the section.

Standard potential

Pungor and Tóth[1] have emphasized the importance of the defect structure of silver halide precipitates used in the preparation of halide electrodes. Later Marton and Pungor[8] proposed a theory relating the standard potential of silver halide-based electrodes to the stoichiometry of the active material; they were able to use the theory to explain changes in the standard potential due to the small variations in the stoichiometry of the materials caused by subjecting them to ultraviolet radiation. Recently Buck and Shepard[9] in a well-argued paper developed a more complete theory, on the same principles, which successfully accounts for the differences in standard potential between the solid state electrodes, in which electron exchange occurs at the contact interface, and the two other types of electrode in this category, electrodes with an internal electrolyte and electrodes of the second kind, in which ion exchange occurs at the contact interface.

The theory of Buck and Shepard[9] is based on consideration of the stoichiometry of the active material in the electrode membrane. The formula of any silver halide may be precisely written $Ag_{1+\delta}X$; thus, the compound is slightly nonstoichiometric and may, while at equilibrium, contain a uniform excess of elemental silver (δ positive) or halogen X (δ negative). For silver bromide, for example, it has been shown[10] that δ may be in the range $-5 \times 10^{-8} \leqslant \delta \leqslant 5 \times 10^{-13}$. Thus, when the mole fraction of excess silver reaches 5×10^{-13}, the silver bromide is saturated with silver and the activity of silver in the compound is unity: similarly the activity of bromine is unity when the mole fraction of excess bromine reaches 5×10^{-8}. These defect elements are in equilibrium with holes and electrons in the crystal structure, and thus the activities of the defect elements will affect the potential owing to electron exchange at the contact interface of the membrane inside the electrode. The same argument may be applied to silver sulfide membranes.

For electrodes of the second kind nonstoichiometry is unimportant, as the silver halide takes no important part in the potential-determining processes: it serves only to ensure that optimum conditions prevail at the electrode surface for the solubility equilibrium between the halide ions and silver ions to be established. The potential of the silver is established as the result of the electron exchange between the silver metal and ions in the bulk of the metal and ion exchange between the silver ions in the metal and the silver ions at the silver–solution interface, as with an electrode of the first kind. Clearly in this situation the activity of silver at the interface is unity and not subject to variation; hence, the standard potential of the electrode is set. The solubility equilibrium on the electrode surface may be written

$$_i a_{Ag^+}\,_s a_{X^-} \rightleftharpoons K_{SO}(AgX) \tag{5.1}$$

where $_i a_{Ag^+}$ is the silver ion activity at the silver–solution interface, $_s a_{X^-}$ is the activity of X^- in the bulk of the solution and $K_{SO}(AgX)$ is the solubility product of AgX. Hence, it may be deduced that the potential of the cell including the electrode of the second kind is given by

$$E = E^\circ{}_{Ag^+/Ag} + \frac{RT}{F} \ln K_{SO}(AgX) - \frac{RT}{F} \ln {}_s a_{X^-} - E_{ref}. \tag{5.2}$$

where E_{ref} is the potential of the reference electrode and $E°_{Ag^+/Ag}$ is the standard potential of a silver electrode (of the first kind). This equation has been experimentally verified on many occasions. The analogous equation for the response of the electrodes to silver ions is

$$E = E°_{Ag^+/Ag} + \frac{RT}{F} \ln {}_sa_{Ag^+} - E_{ref}.$$ (5.3)

hence, the potential is independent of the silver salt.

In the case of electrodes with an internal reference solution and a silver-based internal reference electrode, theoretical analysis leads to the same equations (Eqs 5.2 and 5.3), since the potentials due to the internal reference solution cancel out, leaving the total of the interfacial potentials in the cell the same as for electrodes of the second kind. All the interfacial potentials result from ion-exchange processes; the only potential due to electron exchange results from the reaction $Ag \rightleftharpoons Ag^+ + e^-$ within the surface of the silver reference electrode. The same equation is obeyed by the solid state halide electrodes contacted inside with silver. Whatever the initial stoichiometry of the AgX membrane before contact was made, when the membrane is contacted with the silver, silver diffuses into the membrane until the stoichiometry has been altered to give unit silver activity. Any excess halogen reacts with the silver until this situation is reached; only then are these two phases in the system at equilibrium. Hence, electrodes of the second kind and the corresponding electrodes with internal reference solutions or solid silver contacts should and do all have the same standard potentials. Nearly all commercial electrodes fall into one of these categories.

If, in a solid state electrode, the membrane is not contacted with elemental silver but with another, noble, metal, the silver activity in the membrane is fixed by the stoichiometry of the silver salt and electron exchange replaces ion exchange as the process responsible for the potential at the membrane–contact interface. The standard potential will thus vary according to the stoichiometry between the limits set by unit activity of silver and unit activity of halogen: the range of standard potentials for the different silver salt membranes is listed in Table 5.2, on the assumption that the membranes are used for measuring silver activity (i.e. the standard potential is the electrode potential when ${}_sa_{Ag^+} = 1$). These values were calculated by Buck and Shepard[9] by substitution into the following equation for the cell potential:

$$E = E°_{Ag^+/Ag} + \frac{RT}{F} \ln {}_sa_{Ag^+} - \frac{RT}{F} \ln a*_{Ag} - E_{ref}.$$ (5.4)

where $a*_{Ag}$ is the activity of the free metal within the membrane referred to a standard state of unit activity of pure silver.

Membranes contacted with carbon, as in the electrodes developed by Růžička and his colleagues,[11] or mercury, as in the experiments of Marton and Pungor,[8]

TABLE 5.2
Extreme values for standard potentials of membranes[9]

	Standard potential $(_sa_{Ag^+} = 1)/V$	
Membrane	$a^*_{Ag} = 1$	$a^*_X = 1$
AgCl	0.799	1.935
AgBr	0.799	1.793
AgI	0.799	1.486
Ag_2S	0.799	1.006
AgSCN	0.799	1.478
	Standard potential $(_sa_{M^{2+}} = 1)/V$	
Membrane	$a^*_{Ag} = 1$	$a^*_s = 1$
PbS/Ag_2S	0.151	0.354
CdS/Ag_2S	0.123	0.326
Cu_xS/Ag_2S	0.448 (Ref. 12)	0.595

show the behaviour expected by this theory, with the standard potential varying according to the method of manufacture and, hence, the stoichiometry of the active material in the membrane. Buck and Shepard produced a range of silver bromide membranes ranging from one extreme of stoichiometry to the other and showed that the standard potentials of the electrodes produced from these membranes agreed well with their theory. A membrane saturated with bromine and contacted with carbon gave an electrode with a standard potential of 1.77 V, whereas a membrane contacted with silver gave 0.797 ± 0.003 V. Membranes contacted with mercury have been shown[8,9] to behave slightly differently when saturated with halogen, because the halogen reacts with the mercury until an equilibrium position is reached between the mercury, mercurous halide, silver halide and halogen; thus, the range of $E°$ values available for mercury-contacted membranes is more restricted than that shown in Table 5.2.

When a membrane is contacted by a reactive metal such as lead, cadmium or solder, the potential is initially similar to that of membranes contacted by a noble metal and only electron exchange occurs at the membrane–contact interface. However, the membrane, AgX, reacts with the contact metal, L, to form salts L_xX_y and free elemental silver. This silver will eventually act as contact metal itself and the standard potential will shift, as the membrane stoichiometry shifts, to that expected for a silver contact.

Buck and Shepard[9] extended their arguments from the silver halide and silver sulfide membranes to cover the mixed sulfide membranes used in lead, cadmium and copper electrodes (PbS/Ag_2S, CdS/Ag_2S, Cu_xS/Ag_2S). These membranes have been considered in more detail in a paper by Koebel,[12] who made particular comment on the copper-selective membrane. As with the electrodes previously discussed, lead, cadmium or copper cannot be used as contact metals because of corrosion caused by the silver salts. Hence, the range of standard potentials is set as before by the limits of stoichiometry when $a^*_{Ag} = 1$ or $a^*_S = 1$: this range, referred to the standard state of $a_{M^{2+}} = 1$ (where M^{2+} is the primary ion, Pb^{2+}, Cd^{2+} or Cu^{2+}) is given by theory as

$$E^\circ_{Ag^+/Ag} + \frac{RT}{2F} \ln K_{SO}(Ag_2S)/K_{SO}(MS) \quad \text{to} \quad E_{M^{2+}/M} + \frac{RT}{2F} \ln {}_{MS}K_0$$

where ${}_{MS}K_0 = a_{MS}/a_M a_S$ (i.e. the stability constant for the reaction $M + S \rightleftharpoons MS$). The values corresponding to these expressions for the different electrodes are given in Table 5.2. These simply derived equations are sufficient to give fair agreement with experimental results, especially for the lead and cadmium electrodes, as demonstrated by Koebel.[12] The equilibria in the copper sulfide/silver sulfide system are much more complicated than in the other two systems because of the range of copper sulfides of grossly different stoichiometry (CuS, $Cu_{1.8}S$, $Cu_{1.9}S$, $Cu_{1.95}S$ and Cu_2S) which can exist. The value of the electrode standard potential ($a_{Cu^{2+}} = 1$) depends on which of the sulfides are present at equilibrium and whether or not the system is in fact at equilibrium. In the extreme case of the membrane being saturated with sulfur ($a^*_S = 1$), the equilibrium position is well defined and the form of copper sulfide present is CuS. The opposite extreme is obtained when the membrane is contacted with silver ($a^*_{Ag} = 1$). The silver is in equilibrium with sulfur in the system via the silver sulfide; thus, the chemical potential of the sulfur and, hence, the chemical potential of the copper/sulfur system is fixed. Analysis of the available thermodynamic data shows that the copper sulfide in equilibrium with the silver-contacted membrane has the formula $Cu_{2+\delta}S$. Unfortunately, the experimental results do not agree very well with theory, probably because the rate of approach to equilibrium of the system is too slow. Most commercial electrodes have silver contacts.

The function of the silver sulfide in these lead, cadmium and copper electrodes has often been discussed. Koebel[12] suggests that the prime function is to help to keep the activity of the metal, M, constant at the membrane–solution interface. This is necessary because the effect of redox reactions on the surface, and also the effects of the small currents drawn by the measurement system through the cell, is a tendency for metal activities in the surface to change and, hence, change the standard potential of the electrode; this is seen by the user as drift. Silver sulfide is unusual in that the silver or sulfur in the solid will diffuse relatively fast. Hence, in any reaction on the surface it tends to act as a buffer in the equilibria involved, with any silver which is lost from or gained on the surface being rapidly equilibrated with the bulk of the membrane by diffusion processes.

Response range

The most comprehensive theoretical treatment of the selectivity and detection limits of the electrodes with membranes composed wholly of silver compounds is that published by Morf, Kahr and Simon.[13] Their theory is supported both by their own experiments, reported in the paper, and by data previously published by other authors.

According to this theory, the lower limit of response is governed both by the dissolution of the membrane to yield small activities of the membrane components, according to the equation

$$K_{SO}(Ag_zX) = {_i}a^z_{Ag^+} {_i}a_{X^{z-}} \tag{5.5}$$

and by the activity of silver ion defects in the membrane. The defects are largely Frenkel defects occupying interstitial sites. If the silver ion defect activity in the silver compound Ag_zX is α mol kg^{-1}, then the following electroneutrality equation is obeyed at the membrane–solution interface:

$$_ia_{Ag^+} - {_s}a_{Ag^+} - \alpha = z({_i}a_{X^{z-}} - {_s}a_{X^{z-}}) \tag{5.6}$$

when the membrane is immersed in a solution. Starting from this general equation, the effects of the defect activity on the response of the electrodes to either silver ions or anions may be deduced.

The response of the electrodes to silver ions is strongly dependent on the relative magnitudes of the defect activity and the solubility product of the membrane material, $K_{SO}(Ag_zX)$. Using the assumption that $_sa_{X^{z-}} = 0$, and substituting Eq. (5.5) into Eq. (5.6) leads to the equation

$$_ia_{Ag^+}^{z+1} - {_i}a_{Ag^+}^{z}({_s}a_{Ag^+} + \alpha) - zK_{SO}(Ag_zX) = 0 \tag{5.7}$$

However, the potential of the electrode can be simply shown to respond according to the equation

$$E = E^{\circ}_{Ag^+/Ag} + \frac{RT}{F} \ln {_i}a_{Ag^+} \tag{5.8}$$

Therefore, the response of the electrode to silver ions in the bulk of the solution may be calculated by substitution of the value of $_ia_{Ag^+}$ from Eq. (5.7) into Eq. (5.8). This is most readily done if certain further assumptions are made to account for the properties—in particular, values of $K_{SO}(Ag_zX)$ and α (see Table 5.3)—of the silver compounds used as membrane materials.

TABLE 5.3
Values of K_{SO} and α for silver compounds[13] at 25 °C

Compound	Solubility product, $-\log_{10} K_{SO}$	Silver ion defect activity, $-\log_{10} \alpha$
AgCl	9.8	8.8
AgBr	12.3	6.3
AgI	16.1	6.0 (estimated value)
Ag$_2$S	48.5^a	5.5 (estimated value)
AgSCN	11.9	

^{a} As shown by the values in Ref. 15 and Table 5.4, this value is uncertain.

For a membrane material such as silver chloride $z = 1$ and $K_{so}(AgX) \gg \alpha_i^2$; in this case Eqs (5.7) and (5.8) reduce to

$$E = E°_{Ag^+/Ag} + \frac{RT}{F} \ln \frac{{}_sa_{Ag^+} + {}_sa_{Ag^+}{}^2 4K_{so}(AgX)}{2}$$ (5.9)

This equation has been produced by several workers and has been shown to describe closely the experimental behaviour of the electrode as shown in Fig. 5.2. The equation predicts that the electrode will give Nernstian response as long as ${}_sa_{Ag^+} \gg (4K_{so}(AgX))^{\frac{1}{2}}$.

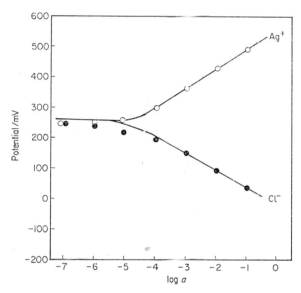

Fig. 5.2 Response of a silver chloride-based ion-selective electrode to silver ions and chloride ions[13]

At the other extreme, in the case of membrane materials such as silver iodide or silver sulfide, $\alpha^{z+1} \gg zK_{so}(Ag_zX)$, and, hence, Eqs (5.7) and (5.8) simplify to

$$E = E°_{Ag^+/Ag} + \frac{RT}{F} \ln ({}_sa_{Ag^+} + \alpha)$$ (5.10)

This behaviour, with Nernstian response only when ${}_sa_{Ag^+} \gg \alpha$, may be illustrated by the response curve of a silver sulfide electrode (Fig. 5.3). The detection limit of the electrode is now α mol kg^{-1}.

The behaviour of an electrode with a silver bromide membrane in silver ion samples lies between these two extremes.

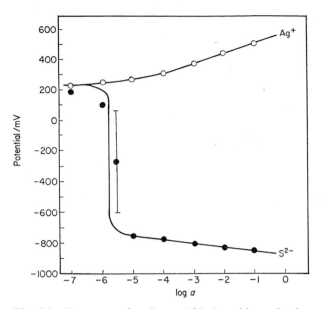

Fig. 5.3   Response of a silver sulfide-based ion-selective
electrode to silver ions and sulfide ions[13]

The response of the electrodes with Ag_zX membranes to the anions X may be calculated analogously from Eq. (5.6). The same two extreme cases may conveniently be treated. For the first case (e.g. AgCl membranes), in which the silver ion defect activity is relatively low, the response obeys Eq. (5.11), as shown in Fig. 5.2:

$$E = E^°_{X_2/X^-} - \frac{RT}{F} \ln \frac{{}_s a_{X^-} + [{}_s a_{X^-}{}^2 + 4K_{SO}(AgX)]^{\frac{1}{2}}}{2} \qquad (5.11)$$

The electrode response in the other extreme case (AgI and Ag_2S membranes), in which the defect activity is relatively high, is not so straightforward and the theory predicts a surprising jump in response at a determinand activity of α/z mol kg^{-1}. Morf et al.[13] managed by careful experimentation to demonstrate this effect (Fig. 5.3).

According to Morf et al., in calculations on those membranes composed of mixtures of silver halides and silver sulfide, the silver ion defect activity of the silver sulfide should be used.

The response range of electrodes in buffered solutions of determinand is markedly different from the response in the unbuffered solutions discussed so far; it appears that the buffering action at the membrane surface of the excess of the buffer solution serves to make the effect of the defect activity negligible. Durst[14] has demonstrated that in solutions of buffered silver activity the response of a silver sulfide electrode is Nernstian down to pAg 15.5, well below the Nernstian limit indicated by Eq. (5.10): such measurements are impossible in unbuffered solutions.

The response ranges of the lead, cadmium and copper electrodes have not been treated as rigorously as the silver halide and silver sulfide electrodes. For the mixed metal sulfide/silver sulfide membranes contacted with silver, the electrode potential may be represented approximately by[9]

$$E = E°_{Ag^+/Ag} + \frac{RT}{2F} \ln (K_{SO}(Ag_2S)/K_{SO}(MS)) + \frac{RT}{2F} \ln {}_sa_M{}^{2+} \quad (5.12)$$

Hence, the simple picture is roughly true of the membrane as a silver sensor with the solubility equilibria of the silver sulfide/metal sulfide system relating the silver ion activity to the metal ion activity at the membrane surface. The activities of silver ions which are sensed at the surface are extremely low ($<10^{-8}$ M) and the equilibria at the interface are effectively buffered; hence, no limit of detection can be deduced from the expected performance of the silver-sensing system as previously outlined; but in solutions buffered with respect to the metal ions the detection limit appears to be set by the solubility of the silver sulfide in the membrane. According to the solubility equilibria, the activity of silver ions on the membrane surface is given by

$$_ia_{Ag^+} = [_ia_M{}^{2+} \times K_{SO}(Ag_2S)/K_{SO}(MS)]^{\frac{1}{2}} \quad (5.13)$$

However, in the absence of silver ions generated because of depression of the sulfide ion activity on the interface, the activity of silver ions settles to the value set by the solubility of silver sulfide, about 10^{-17} M. Substitution of this value and the appropriate solubility products (see Table 5.4) into Eq. (5.13) leads to a limit of detection of about 10^{-11} M for the lead and cadmium electrodes; these figures agree satisfactorily with experimental results (see section on response ranges). The case of copper sulfide is more difficult because of uncertainty in the solubility product of the form of copper sulfide present. The limits of detection of these and other heavy metal electrodes in unbuffered solutions are presumably set by the parameters discussed in Chapter 3.

Selectivity

The interferences with electrode response caused by cations, anions and ligands are to be discussed separately.

Cation interference

Very few cations react with the silver compounds used for membrane preparation and, hence, interference caused by cations is seldom a problem. However, one cation which does react with the compounds and interfere with electrode response is the mercuric ion, Hg^{2+}. The nature of the interference is different for the different membranes, as it depends in part on the properties of the mercuric salt formed on the membrane surface. On the surface of silver sulfide electrodes mercuric ions form a blocking film of mercuric sulfide which rapidly destroys the response, but on the halide electrodes mercuric ions form soluble halide complexes and liberate silver ions. In the case of the iodide electrode, the potentials resulting from the sensing of these silver ions are sufficiently stable and reproducible for the electrode to be used as a sensor for mercuric ions, albeit over a narrow concentration range. The reaction on the electrode surface is[13]

$$Hg^{2+} + nAgI \rightleftharpoons HgI_n^{2-n} + nAg^+ \qquad (5.14)$$

At low mercuric ion concentrations ($< 10^{-4}$ M) $n \simeq 1$, and a slope approaching 59 mV per decade is obtained; however, at higher concentrations n tends towards 2, with a resultant slope of approximately 29.5 mV per decade.

The interference of heavy metal ions with the silver sulfide-based lead, cadmium and copper electrodes is similar to the interference of mercuric ions with the silver sulfide-based silver electrode. Any metal ion forming a more insoluble sulfide than the sulfide of the determinand will interfere with the electrode response. Solubility products of the relevant sulfides are given in Table 5.4; there is a wide span of values given for each compound in the tables[15]

TABLE 5.4
Solubility products of some heavy metal sulfides at 25 °C[15]

HgS	10^{-52}	CdS	10^{-28}
Ag_2S	10^{-51}	PbS	10^{-28}
CuS	10^{-36}	Bi_2S_3	10^{-96}

and, hence, the values in Table 5.4 should be treated as approximations. Hence, mercury and silver ions are expected to interfere with all three electrodes, even if present at very low levels, but, fortunately, when the interferent concentration is less than 10^{-7} M, the rate at which the interfering ions diffuse to the membrane surface and react with the membrane material is too slow for interference to be observed. At higher concentrations of interferent (10^{-7}–10^{-5} M) the interferent, J^{2+}, reacts with the membrane surface according to the reaction

$$MS + J^{2+} \rightleftharpoons JS + M^{2+} \qquad (5.15)$$

Interference is then observed if the amount of M^{2+} released according to Eq. (5.15) is sufficient to increase detectably the concentration of M^{2+} in the bulk of the sample. The accumulation of JS on the membrane surface may eventually poison the electrode and a new membrane surface will need to be exposed by abrasion. It is clear from the mechanism of the interference that it is immaterial which ion is interfering, except in so far as it affects the rate of reaction, but the important point is only that the equilibrium expressed by Eq. (5.15) has a tendency to move to the right. Thus, taking the cadmium electrode as an example, the interferences caused by low levels of silver, mercuric or cupric ions are identical.[16] At higher concentrations of interferent the electrodes act as pure silver sulfide electrodes and will, for example, give a Nernstian response to silver ions. In general, it is advisable when using these electrodes to keep the activity of heavy metal interferents below 10^{-6} M.

Anion interference

The halide and sulfide electrodes suffer interference from ions forming insoluble silver salts. In particular, interference is experienced if the sample of determinand contains anions which form silver salts more insoluble than that formed by the determinand.

Potential interferents include halides, sulfide, cyanide, thiocyanate and hydroxide. If for a membrane material AgX and an interferent Y^{n-} the concentration is sufficiently high to push to the right the equilibrium:

$$nAgX + Y^{n-} \rightleftharpoons Ag_nY + nX^- \tag{5.16}$$

then the compound Ag_nY will be formed on the membrane surface. This compound, depending upon the identity of X and Y, may either form a mixed solid phase with AgX or form a film covering the surface of AgX: the electrode response is different in the two cases.[13]

When a mixed phase is formed, as, for example, when bromide ions in a sample of chloride ions react with the membrane to form an AgCl + AgBr phase, the electrode responds according to the straightforward equation

$$E = E^{\circ}{}_{X_2/X^-} - \frac{RT}{F} \ln (_s a_{X^-} + k_{XY} \, _s a_{Y^{n-}}) \tag{5.17}$$

where

$$k_{XY} = \frac{K_{SO}(AgX)}{K_{SO}(Ag_nY)^{1/n}}$$

In this situation both ions contribute to the electrode behaviour; a typical response curve is shown in Fig. 5.4 (curve A).

The alternative behaviour, when Ag_nY forms a complete film on the surface of AgX, is seen when X^- is Br^- and Y^{n-} is SCN^-. If, in the sample

$$a_{X^-} < \frac{K_{SO}(AgX)}{K_{SO}(Ag_nY)^{1/n}} \times a_{Y^{n-}}{}^{1/n} \tag{5.18}$$

then the electrode response obeys the equation

$$E = E^{\circ}{}_{Y/Y^{n-}} - \frac{RT}{nF} \ln \, _s a_{Y^{n-}} \tag{5.19}$$

Conversely, if

$$a_{X^-} > \frac{K_{SO}(AgX)}{K_{SO}(Ag_nY)^{1/n}} \times \alpha_{Y^{n-}}{}^{1/n} \tag{5.20}$$

the electrode response is as if no Y^{n-} were present at all

$$E = E^{\circ}{}_{X_2/X^-} - \frac{RT}{F} \ln \, _s a_{X^-} \tag{5.21}$$

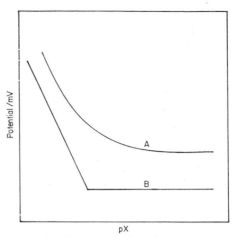

Fig. 5.4 Curves showing the response of electrodes based on AgX to varying concentrations of the primary ion, X, in the presence of a constant background of interferent, Y, when (A) AgX and AgY form a mixed phase; (B) a single phase of either AgX or AgY completely covers the AgX membrane

Thus, the electrode responds to either X^- or Y^{n-} according to their relative concentrations, but not to both at once. The shape of a typical response curve of this type is also included in Fig. 5.4 (curve B).

The agreement between calculated and experimental values of k_{XY} is remarkably good, as shown in Fig. 5.5.

Chloride and bromide ions interfere seriously with the response of the copper electrode based on copper sulfide/silver sulfide. It has been suggested by Ross[17] that the chloride interference, for example, is due to the reaction on the membrane surface:

$$Ag_2S + Cu^{2+}(aq) + 2Cl^- \rightleftharpoons 2AgCl + CuS \qquad (5.22)$$

Crombie, Moody and Thomas[18] supported this with evidence that copper electrodes exposed to large concentrations of chloride ions afterwards responded in an approximately Nernstian manner to chloride ions, thus implying the formation of the silver chloride layer. However, chloride ions also complex cupric ions, and thus the relationship between the cupric ion concentration and the chloride ion concentration is not as simple as that presented in Eq. (5.22). The experiments of Crombie et al. showed that the contributions to the electrode response from cupric ions and chloride ions were equal if

$$2.8pCl = pCu + 2.7 \qquad (5.23)$$

The bromide interference is approximately 10^3 times greater than the chloride interference.

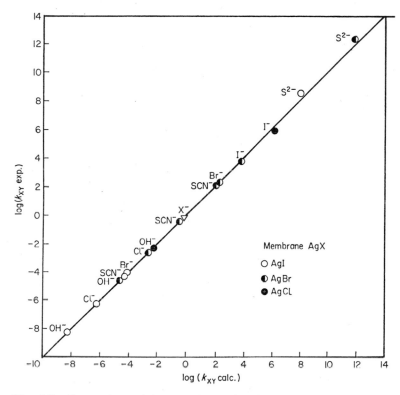

Fig. 5.5 Comparison of the experimental and calculated values of k_{XY} for electrodes based on the silver halides[13]

Ligand interference

The interference caused by ligands is due to reactions of the type

$$AgX + zL^{n-} \rightleftharpoons AgL_z^{1-nz} + X^-$$ (5.24)

The electrode senses the X^- ions generated at the membrane–solution interface in addition to any X^- ions from the bulk of the solution. The position of the equilibrium in reaction (5.24), and, hence, the activity of X^- generated, depends not only on the activity of L^{n-} but also on the rate of diffusion of L^{n-} ions to the interface, to replace those consumed by the reaction, and on the rate of diffusion of AgL^{1-nz} ions from the membrane. A full treatment of the response of electrodes of the second kind to ligands, which takes account of these parameters and more, was presented several years ago by Jaenicke and reviewed more recently by Vetter.[19] The equations for the responses of the electrodes are complex but may be simplified in certain cases. Two reasonable approximations are to assume that $z = 2$ (since Ag^+ has a tendency to form linear $1:2$ complexes) and that the two diffusion rates mentioned above are equal: the response of the electrodes in solutions containing either L^{n-} species only or both L^{n-} and X^- species may then be considered.

The case of the solutions to be analysed containing only L^{n-} species is particularly interesting, as in such solutions the electrode response is entirely due to L^{n-}. Morf et al.[13] have deduced that the response in such conditions is

$$E = E^{\circ}_{X_2/X^-} - \frac{RT}{F} \ln \frac{2\kappa_s a_{L^{n-}}}{4\kappa_s a_{L^{n-}} + \alpha + (4\kappa_s a_{L^{n-}}^2 + \alpha^2)^{1/2}} \quad (5.25)$$

where $\kappa = \beta_2 K_{SO}(AgX)$ (where β_2 is the formation constant $_s a_{AgL_2}^{1-2n}/$ $_s a_{L^{n-}}^2 \times {}_s a_{Ag^+}$) or $\beta_2 K_{SO}(Ag_2S)$ for Ag_2S membranes. They plotted values of $0.5 \log \kappa$ against $E - E^{\circ}_{X_2/X^-}$ for different ligand activities with different ligands and different membrane materials. Figure 5.6 shows the graph[13] obtained for $_s a_{L^{n-}} = 10^{-2}$ M; it may be seen that the curve has three distinct regions.

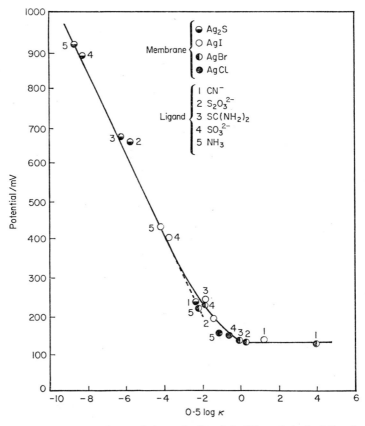

Fig. 5.6 Comparison of the calculated (solid and dashed lines) and experimental responses of various membranes to various ligands, all at a concentration of 10^{-2} M[13]

On the right of the graph, where $4\kappa \gg 1$, AgL_2^{1-2n} species predominate at the membrane–solution interface; Eq. (5.25) reduces to

$$E = E^°_{X_2/X^-} - \frac{RT}{F} \ln (\tfrac{1}{2s}a_{L^{n-}})$$ (5.26)

Thus, for example, the iodide electrode responds to and may be used for measuring cyanide ions with a slope of 59 mV decade^{-1} at 25 °C, but the standard potential is shifted by 18 mV with respect to the standard potential for iodide measurements. The equation is no longer obeyed if the cyanide activity is sufficiently large for appreciable formation of $Ag(CN)_3^{2-}$ to occur at the interface.

In the central part of the curve in Fig. 5.6, where $1 \gg 4\kappa \gg (\alpha/_s a_{L^{n-}})^2$, free ligands predominate at the interface but $_i a_{X^-}$ and $_i a_{AgL}^{1-2n}$ are greater than α; Eq. (5.25) reduces to

$$E = E^°_{X_2/X^-} + \frac{RT}{F} \ln (\kappa^{\frac{1}{2}}_s a_{L^n})$$ (5.27)

The response of the silver bromide electrode to sulfite ions is an example of this behaviour.

The final region to the left of the curve is the most curious. In this region the behaviour of the most insoluble membranes in the presence of ligands is shown; dissolution of the membranes is very slow. Here $4\kappa \ll (\alpha/_s a_{L^{n-}})^2$ and, hence,

$$E = E^°_{X_2/X^-} + \frac{RT}{F} \ln \alpha - \frac{2RT}{F} \ln (\kappa^{\frac{1}{2}}_s a_{L^{n-}})$$ (5.28)

The apparent standard potential of the electrode for the measurement of ligand activity is thus dependent on the silver ion defect activity, α, and the electrode response is produced by reaction of the ligand with these interstitial silver ions. The response slope of the electrode to the ligand is 118 mV decade^{-1} at 25 °C. Thus, for example, a silver sulfide electrode will respond to cyanide ions or ammonia with a slope of 118 mV decade^{-1}. This cyanide response has been shown to be of some practical use: the selectivity of the response may be roughly assessed from Fig. 5.6. The stability of these responses is, unfortunately, not always satisfactory and response times tend to be long.

If the solutions to be analysed contain both L^{n-} and X^- species, the electrode response has contributions from both. When $\kappa \gg 1$, in the case of relatively strong silver-ligand complexes, Morf et al.[13] have shown that the equation for the electrode response is

$$E = E^°_{X_2/X^-} - \frac{RT}{F} \ln (_s a_{X^-} + \tfrac{1}{2s}a_{L^{n-}})$$ (5.29)

or, more precisely, taking into account the ionic diffusion rates, D, of ions to and from the membrane,

$$E = E^°_{X_2/X^-} - \frac{RT}{F} \ln \left(_s a_{X^-} + \frac{D_{L^{n-}}}{2D_{X^-}} _s a_{L^{n-}}\right)$$ (5.30)

This equation describes, for example, the response of the silver iodide electrode in a sample containing both iodide and cyanide ions: the selectivity coefficient k_{ICN} is thus equal to $D_{Ln^-}/2D_{X^-}$ or, very approximately, $\frac{1}{2}$. The opposite extreme of behaviour in mixed solutions, when $\kappa \ll 1$, leads to the equation

$$E = E^\circ{}_{X_2/X^-} - \frac{RT}{F} \ln \left({}_s a_{X^-} + \kappa \frac{D_{AgL_2}{}^{1-2n}}{D_{X^-}} \times \frac{{}_s a^2{}_{L^{n-}}}{{}_s a_{X^-}} \right) \qquad (5.31)$$

Therefore, in this case the selectivity coefficient is not a constant but is dependent on the activities of both the X^- ions and the ligand.

The theory of the operation of electrodes of the second kind, much of which is directly applicable to ion-selective electrodes, has been reviewed thoroughly by Vetter.[19] He considers the various mechanistic paths which may lead to the response and also the effect of the conductivity of the electrode coating. Electrode behaviour is discussed in relation to the dissolution current density of the membrane, the exchange current density at the membrane–solution interface and the charge transfer overvoltage.

Lanthanum fluoride electrode

The response mechanism of the lanthanum fluoride electrode is apparently less complicated than those of the electrodes already discussed. It has been assumed, since soon after the announcement of the electrode, that the limit of detection was fixed by the solubility of the membrane (e.g. Ref. 17), as in the case of the electrodes based on silver chloride. This has more recently been supported by the experiments of Buffle, Parthasarathy and Haerdi,[20] who concluded that the two primary factors which limit the electrode sensitivity are the solubility of the membrane and the adsorption of fluoride ions on to the membrane surface.

Durst and Ross[21] coulometrically generated fluoride ions from a lanthanum fluoride membrane with an efficiency of $99.2 \pm 0.5\%$; this suggested that fluoride ions are the only mobile species in the crystal and are wholly responsible for the carriage of charge. Stahr and Clardy[7] also concluded, from experiments in which they studied the migration of F^{18} anions into the membrane, that the mechanism of charge conduction was via the fluoride ions. These authors also confirmed the proposal of Brand and Rechnitz[5] that a film, possibly of lanthanum hydroxide, is present on the membrane surface and participates in the response mechanism.

PERFORMANCE

Response range and slope

The response ranges of the anion-selective electrodes which are commercially available are given in Table 5.5: data are taken, where possible, from independent evaluations rather than manufacturers' publicity material. No differentiation is made between the performance of the different commercial electrodes for the same determinand, as virtually all function in an essentially identical manner and, hence, the response ranges and slopes are very similar except where stated. The performances of the recently announced electrodes of Sekerka and Lechner[22,23] are included for comparison. The ranges of pH values within which samples can be measured with the electrodes are also included in Table 5.5.

TABLE 5.5
Response ranges of anion-selective electrodes at 25 °C

Electrode	Active membrane material	Upper useful limit/M	Limit of Nernstian response[a] M and (mg l^{-1})		Limit of detection[a]/M	pH range
F$^-$	LaF$_3$	sat. solns	2×10^{-6}	(0.04)	10^{-7}	5–8
Cl$^-$	AgCl	1	10^{-4}	(3)	10^{-5}	2–11
	Hg$_2$Cl$_2$	—	2×10^{-6}	(0.07)	5×10^{-7}	0–6
Br$^-$	AgBr	1	10^{-5}	(0.8)	10^{-6}	2–12
	Hg$_2$Br$_2$	—	10^{-6}	(0.08)	10^{-7}	1–6
SCN$^-$	AgSCN	1	5×10^{-5}	(3)	5×10^{-6}	2–12
	Hg$_2$(SCN)$_2$	—	5×10^{-6}	(0.3)	5×10^{-7}	1–6
I$^-$	AgI	0.2	2×10^{-6}	(0.25)	10^{-8}	3–12
CN$^-$	AgI	10^{-2}	10^{-5}	(0.25)	10^{-6}	11–13
S^{2-}	Ag$_2$S	sat. solns	5×10^{-6}	(0.15)	10^{-7}	13–14

[a] The response limits and pH ranges refer to solutions unbuffered with respect to the determinands. The definitions of the terms used are those given in Chapter 3.

The upper concentration limits for the halide electrodes result from the onset of membrane dissolution. For example, when a silver chloride membrane is immersed in a strong chloride solution, the membrane slowly dissolves in the solution with formation of polychloroargentate(I) species, primarily $AgCl_2^-$ and $AgCl_3^{2-}$, and the response becomes non-Nernstian and drifts. If, as in the case of the silver/silver chloride reference electrode, the solution in contact with the electrode is saturated with silver chloride, stable readings are obtained; otherwise the electrode potential drifts until sufficient silver chloride has dissolved for equilibrium to be reached, and the electrode surface becomes pitted and the response unstable. This behaviour is particularly marked in the case of the iodide electrode.[24,25] For the cyanide electrode the upper limit is also set by membrane dissolution. The response mechanism involves the dissolution of the silver iodide and this occurs at a rate dependent on the cyanide concentration; when this concentration exceeds 10^{-2} M, the membrane lifetime becomes impractically short, and for continuous use it is advisable to ensure that the concentration does not regularly exceed 10^{-3} M; above this concentration, formation of a significant concentration of $Ag(CN)_3^{2-}$ ions in addition to $Ag(CN)_2^-$ ions leads to a non-Nernstian response.[26]

The lower limits of detection of the electrodes based on silver salts usually agree well with those predicted by the theory outlined in the previous section. However, occasionally differences are apparent between experimental data and theory; in particular, there have been several reports of the use of sulfide electrodes below the break in the curve resulting from the defect activity, even down to 10^{-7} M.[27] This apparently non-theoretical behaviour is only shown by some manufacturers' electrodes, usually when new; after a period of time, often several months but sometimes a year or two, these electrodes seem to revert to the theoretical behaviour. Presumably the method of manufacture of these electrodes is such that the silver defect activity in the silver sulfide is minimized, but it may be that these membranes are not completely at equilibrium. However, many new electrodes do show the predicted break in the calibration curve.[28]

The detection limit of the fluoride electrode has been shown by Baumann[29] to be much lower (2×10^{-10} M = pF 9.7) in solutions buffered with respect to fluoride ions than in unbuffered solutions.

The pH ranges quoted are those applicable to direct measurements of the total activity of determinand in solutions of unknown determinand activity and unbuffered with respect to the determinand. The upper pH limits of the fluoride and chloride electrodes are fixed by the onset of hydroxide ion interference when the determinand activity is at the limit of detection, and are thus dependent on k_{FOH} and k_{ClOH}. It follows, therefore, that the pH range is extended at higher determinand activities; at pF 1, the upper pH limit for samples of fluoride is approximately pH 12. The lower pH limits for samples of fluoride, iodide, cyanide and sulfide are the values of pH above which the proportion of determinand present in the undissociated acid form is insignificant; below pH 13, a significant proportion of S^{2-} ions are present as HS^- or H_2S; below pH 11, CN^- is converted to HCN and so forth. These lower limits are independent of the activity of the determinand. If measurements are carried out below the quoted limits, the electrode senses that fraction of the determinand present in the dissociated form; thus, for example, it is quite possible to use the fluoride electrode to analyse samples at pH 2 but only that proportion of the total fluoride species present as undissociated fluoride ions, F^-, will be sensed by the electrode. Below pH 2, the silver salt-based membranes are attacked by hydrogen ions.

All the electrodes behave sufficiently in accord with theory to give virtually Nernstian response—that is, 59 mV per decade for all the electrodes except the sulfide electrode, which gives 29.5 mV per decade, at 25 °C.

The cation-selective electrodes based on inorganic salts respond over similar ranges to the anion-selective electrodes, as shown in Table 5.6. The response limits and the working pH ranges for these electrodes are all similar (except in the case of the mercury response of the silver iodide electrode), as they all follow the pattern illustrated for the lead electrode (Fig. 3.2). The pH range varies with the determinand activity. At low pH the membrane is attacked by hydrogen ions with formation of HS^- and H_2S, and at high pH the metal ions in the sample are complexed by hydroxyl ions. The pH ranges given in Table 5.6 are approximately mid-scale values; for determinand activities in the region of

TABLE 5.6
Response ranges of cation-selective electrodes at 25 °C

Electrode	Upper useful limit/M	Limit of Nernstian response/M and (mg l^{-1})		Limit of detection/M	pH range
Ag$^+$	sat. sols	2×10^{-6}	(0.25)	10^{-7}	2–9
Pb^{2+}	10^{-1}	10^{-5}	(2)	10^{-7}	3–7
Cd^{2+}	10^{-1}	10^{-5}	(1)	10^{-7}	3–7
Cu^{2+}	1	10^{-7}	(0.006)	10^{-9}	3–7
Hg^{2+}	10^{-4}	5×10^{-7}	(0.1)	10^{-8}	4–5

10^{-1}–10^{-2} M the pH range will be wider (e.g. pH 2–7 for the lead electrode) but at 10^{-7} M much narrower (e.g. pH 8–10 for the lead electrode).

In metal buffer solutions the Nernstian range of the electrodes may be greatly extended as interferences from both hydrogen and hydroxyl ions are suppressed. Blum and Fog[30] and Hansen, Lamm and Růžička[31] have both used Cu^{II}/EDTA and Cu^{II}/NTA buffers in experiments with copper electrodes: Hansen *et al.* obtained satisfactory response down to pCu 13. Pick, Tóth and Pungor[32] used the Cu^{II}/ammine system to produce buffers working down to pCu 11.9. Similarly, buffers for Cd^{2+} and Pb^{2+} have been used to demonstrate electrode response down to pCd 11[33] and pPb 11.3.[34]

Selectivity

A large amount of data on the selectivity of the halide and pseudohalide electrodes has been published and, in general, the measured selectivity coefficients agree well with those calculated according to the theory given in the theoretical section (as shown in Fig. 5.5). A summary of selectivity coefficients for the more important electrodes is given in Table 5.7: the values given are derived from the most reasonable values given in the literature. These coefficients should not be so dependent on the conditions of measurement that full details of these conditions need be given, despite the wide divergence of some reported values. Table 5.7 is not a comprehensive list of all interferents, only of the more important of them; any species forming a strong silver complex[1] (e.g. PO$_4^{3-}$, Fe(CN)$_6^{4-}$, SO$_3^{2-}$) must be considered as a potential interferent and the selectivity coefficient derived from theory. Species which change the determinand activity, as distinct from species which themselves give rise to an electrode response, will be considered in the section on applications.

The only important interferent for the fluoride electrode is hydroxyl ion. Ross[17] has stated that citrate ions interfere with the electrode response, but, according to Evans, Moody and Thomas,[36] neither citrate ions nor acetate, formate or oxalate ions interfere.

From consideration of the equilibria, it might be expected that sulfide ions would interfere with the determination of all other determinands even if only the very slightest traces (e.g. 10^{-15} M) were present. However, in practice at such very low concentrations the rate of the reaction of the sulfide ions with the membrane is sufficiently slow and the rate of aerial oxidation of the sulfide ions sufficiently fast for no interference to be observed. It is not until the sulfide ion concentration exceeds 10^{-7} M that any effect is obtained; then the sulfide ions must be destroyed, usually by oxidation.

TABLE 5.7
Selectivity coefficients of anion-selective electrodes at 25 °C

Electrode	Active membrane material	Interferent	Selectivity coefficient
F^-	LaF_3	OH^-	0 (10^{-1} M F^-) to >1 (10^{-5} M F^-)[35]
Cl^-	$AgCl$	Br^-	3×10^2
		I^-	2×10^6
		OH^-	1×10^{-2}
		$S_2O_3^{2-}$	1×10^2
		CN^-, S^{2-}	interfere strongly
	Hg_2Cl_2	Br^-	6×10^2
		I^-	3×10^3
		CNS^-	2
Br^-	$AgBr$	Cl^-	3×10^{-3}
		OH^-	3×10^{-5}
		I^-	5×10^3
		CN^-, S^{2-}	interfere strongly
	Hg_2Br_2	Cl^-	10^{-4}
		I^-	10^6
		SCN^-	10^{-2}
I^-	AgI	Cl^-	4×10^{-7}
		Br^-	2×10^{-4}
		OH^-	10^{-8}
		CN^-	0.5
		S^{2-}	interferes
CN^-	AgI	Cl^-	10^{-6}
		Br^-	2×10^{-4}
		I^-	1.5
S^{2-}	Ag_2S		no interference

Cyanide ions interfere with the responses of the bromide and chloride electrodes by the mechanism described in the theoretical section. It is not realistic to quote values of k_{BrCN} or, particularly, k_{ClCN}, since the electrodes drift too much in cyanide solutions to be useful.

Most of the electrodes show interference curves of type A (Fig. 5.4): the only quoted example of a system giving a type B curve is the interference of thiocyanate ions on the bromide electrode.

Selectivity data for the cation-selective electrodes based on inorganic salts is given in Table 5.8.

The interference of chloride and bromide ions with the response of the copper

TABLE 5.8
Selectivity data for cation-selective electrodes at 25 °C

Electrode	Selectivity data (interferent—selectivity coefficient)	Principal Ref.
Ag^+	no interferents except Hg^{2+}	
Cu^{2+}	no interference from Pb^{2+}, Cd^{2+}, Zn^{2+}, Co^{2+}, Ni^{2+}, Mn^{2+}	37
	Ag^+, Hg^{2+}, Bi^{3+}, Cl^- and Br^- interfere	
Cd^{2+}	$H^+ - 5 \times 10^{-4}$, $Pb^{2+} - 5 \times 10^{-1}$, $Zn^{2+} - 10^{-4}$	38 (separate soln method)
	$Co^{2+} - 5 \times 10^{-5}$, $Ni^{2+} - 5 \times 10^{-6}$	
	$Mg^{2+} - 1.6 \times 10^{-4}$, $Ca^{2+} - 2.2 \times 10^{-4}$,	
	$Zn^{2+} - 4.1 \times 10^{-4}$, $Co^{2+} - 2.0 \times 10^{-2}$,	
	$Ni^{2+} - 3.2 \times 10^{-4}$, $Al^{3+} - 1.3 \times 10^{-1}$	39 (separate soln method)
	$H^+ - 2.4$, $Mn^{2+} - 2.7$, $Pb^{2+} - 6.1$	
	$Tl^+ - 122$, $Fe^{2+} - 196$	
Pb^{2+}	$Cd^{2+} - 3 \times 10^{-1}$, $Mn^{2+} - 3 \times 10^{-4}$,	40
	$Zn^{2+} - 2 \times 10^{-4}$, $Fe^{2+} - 5 \times 10^{-2}$	(separate soln method)
Hg^{2+}	Ag^+ only major interferent	41

electrode has already been discussed in the theoretical section. Figure 5.7 shows the results of Crombie et al.[18] for the effect of different chloride concentrations on the response of a copper electrode. The difference between the different sets of data for the cadmium electrode is surprisingly large even although the two electrodes tested were not identical (Ref. 37: CdS/Ag_2S/polythene electrode; Ref. 38: Orion CdS/Ag_2S electrode).

Oxidants interfere severely with the response of the copper, cadmium and lead electrodes by oxidizing the membrane surface; this eventually causes the response to fail. Johansson and Edström[42] have considered the effect quantitatively for

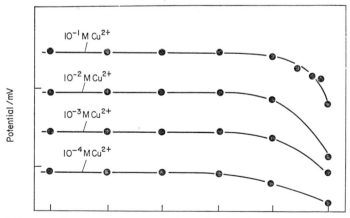

Fig. 5.7 Response of a copper ion-selective electrode to various cupric ion concentrations in the presence of a range of chloride concentrations[18]

the copper electrode. This oxidation effect is one of the tiresome features of the electrodes and strictly limits their application. They are oxidized not only by ions such as Fe^{3+} and $Cr_2O_7^{2-}$ but also, more slowly, by aerial oxygen. Thus, experiments are usually carried out in mildly reducing media, and the electrodes are best stored either absolutely dry or in a solution of a mild reductant (e.g. ascorbic acid). Frequent repolishing of the membrane to uncover a new surface is a feature of the procedures which have to be used in the application of these electrodes.

Stability and reproducibility

The class of electrodes based on inorganic salts includes the most stable and reproducible of all ion-selective electrodes. The fluoride, silver, halide and sulfide electrodes all give very stable and reproducible readings; if used with care, and if the temperature is controlled adequately, there should be no difficulty in achieving stability and reproducibility of ± 0.1 mV. As an example, Srinivasan and Rechnitz[43] found that the fluoride electrode, when used in stirred solutions, gave readings changing by not more than 0.1 mV in an hour and that the standard potential varied by 7 mV in a month without change of slope. After an electrode has been in use for some time, dissolution of part of the membrane surface eventually produces pitting. This adversely affects both the reproducibility and the response time, as small volumes of solution are trapped in the crevices and transferred from one sample to the next. The original performance of the electrode may be restored by polishing the membrane with an inert, fine polishing powder.

The performance of the Orion copper electrode has been carefully studied by Johansson and Edström.[42] They showed, in photomicrographs, the pitting on the surface of an electrode which had been in use for two months: this electrode gave a response slope of 16.2 mV per decade and drifted at a rate of about 1 mV min^{-1} in a 0.01 M copper nitrate solution. After the membrane was polished with diamond paste and treated with silicone oil (to fill residual pits), the electrode gave readings stable to ± 0.01 mV after 2 min in the 0.01 M copper nitrate solution and had a response slope of 29.2 mV per decade at 23 °C, which is very close to the theoretical Nernstian slope. The polishing treatment serves also to remove any oxidized layer on the membrane surface. This treatment was shown to be more effective for the Orion electrode than the EDTA etching procedure recommended by Hansen et al.[31] for their impregnated graphite electrodes; the EDTA treatment produced diffuse pits on the Orion membrane surface and the electrode gave a slightly nonlinear calibration graph. The behaviour of the copper electrode is similar to that of the lead and cadmium electrodes, although the situation is aggravated by the greater solubilities of lead sulfide and cadmium sulfide compared with copper sulfide.

Response times

The responses of all the electrodes based on inorganic salts are very fast and the response speed seldom limits the applications of the electrode except in solutions of concentration very close to the limit of detection. In general, the electrodes respond more quickly in strong or buffered solutions of determinand and in solutions with a high (i.e. 10^{-2}–10^{-1} M) background concentration of inert electrolyte.

Tóth and co-workers[44,45] have studied the response times of halide, cyanide and copper electrodes, using an apparatus designed to prevent the cell going open-circuit when the determinand concentration was changed; the apparatus, which is illustrated in Ref. 45, produces the solution change in less than 1 ms. In all cases the measured response times for one or two decade concentration changes, at concentrations in the middle of the Nernstian response range, were of the order of several hundred milliseconds. The rate of potential change was interpreted by the authors in terms of the rate of desolvation of the determinand ions at the membrane–solution interface. Similar results with halide electrodes have been obtained by Rangarajan and Rechnitz.[46] These response times apply only to such idealized conditions where the concentration change is very rapid, there is a high and steady flow rate of the sample across the membrane surface and the cell circuit remains unbroken: when measurements are taken in beakers, even if the samples are stirred, the response times are by common experience much longer. For decadic concentration changes in the middle of the Nernstian response range, the response time of polished electrodes (to 1 mV from the final equilibrium potential) is typically of the order of 10 s. This time includes not only the response time of the electrode but also the time taken by the liquid junction of the reference electrode to stabilize.

At lower concentrations the response times are longer, particularly in the region between the limit of Nernstian response and the limit of detection. Blaedel and Dinwiddie[47] showed that the response time of a copper electrode is strongly dependent on the copper ion concentration (Fig. 5.8) and also on the

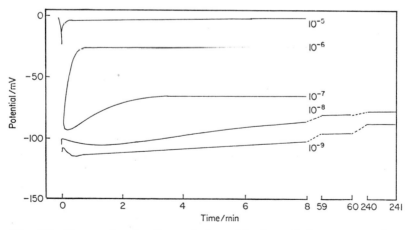

Fig. 5.8 Curves showing the variation with time of the response of a copper-selective electrode to a range of copper ion concentrations[47]

presence of other cations in the sample: potassium ions appear to accelerate the response, whereas aluminium ions decelerate it. In a stirred 10^{-7} M Cu^{2+} solution the response time is approximately 4 min. The results of Smith and Manahan[48] showed rather slower response times (i.e. about 20 min at 10^{-7} M Cu^{2+}, but less than 1 min at 10^{-4} M Cu^{2+}). As an example showing the faster response in metal buffers, Hansen et al.[31] quote the response time of a copper 'Selectrode' in buffers down to pCu 9 as 0.5–1 min.

The fluoride electrode has been widely used in automated analysis systems. Erdmann[49] found that the response of the electrode was fast enough, when total ionic strength adjustment buffer (T.I.S.A.B.) spiked with fluoride was used, to enable samples in the concentration range 0.1–2 mg l^{-1} to be analysed at the rate of 30 per hour in a modified Technicon AutoAnalyzer. For samples in a lower concentration range, 10^{-5}–10^{-7} M (0.2–0.002 mg l^{-1}), Sekerka and Lechner[50] allowed 10 min for the electrodes to reach equilibrium: their automated method thus allowed samples to be analysed at the rate of 6 per hour.

Sensitivity to temperature and light

The temperature coefficients of ion-selective electrodes have been studied by Negus and Light[51] and Lindner, Tóth and Pungor[52]: the results are tabulated in Table 5.9 and compared with the coefficients for the corresponding first- and

TABLE 5.9
Temperature coefficients

Electrode	Nonisothermal temperature coefficient $(dE°/dT)$/mV K^{-1}		
	Ion-selective electrodes		1st and 2nd kind electrodes Ref. 53
	Ref. 51	Ref. 52	
Ag^+	−0.13	−0.06	−0.22
Cl^-	+0.20		+0.21
Br^-	+0.36	+0.45	+0.36
I^-	+0.59	+0.62	+0.59
CN^-	+0.49		
S^{2-}	−0.22	−0.17	−0.21
Cu^{2+}	+1.09	+0.93	+0.88

second-kind electrodes.[53] Negus and Light also report isopotential data for these electrodes. For all the silver-contacted electrodes and electrodes with silver internal reference electrodes, the temperature coefficient should be independent of the construction and similar to the values for the first- and second-kind electrodes: thus, the results of Negus and Light with the halide and silver electrodes and the results of Lindner et al. with the silver and sulfide electrodes are in fair agreement with the coefficients of the first- and second-kind electrodes. The other results of Lindner, with mercury-contacted membranes, and the copper electrode coefficients are dependent on the internal construction of the particular electrode used in the experiments and thus not generally applicable.

Ross, Frant and Riseman[54] have studied the effect of light on membrane pellets formed at a pressure of 10–25 000 psi (70–170 M Pa) from silver chloride and silver sulfide in different proportions; their results on the light effect, together with other properties of the different pellet compositions, are given in Table 5.10. It should be noted that the photoelectric effect observed

TABLE 5.10
Properties of AgCl/Ag$_2$S pellets[53]

No.	Pellet composition mol % AgCl:Ag$_2$S	Volume resistivity/ Ω cm	Photoelectric[a] effect/mV	Limit of Nernstian response to Cl$^-$/M	Other properties
1	100:0	1×10^7	30	10^{-5}	soft, clear
2	90:10	7×10^4	3	10^{-5}	properties between 1 and 3
3	50:50	1×10^4	1.5	10^{-5}	black, hard, dense, imporous
4	10:90	1.5×10^4	1.5	10^{-3}	hard, dense
5	0:100	—	—	no Cl$^-$ response	

[a] Potential change when level of illumination changes from dark to normal room light.

with the pure silver chloride pellet is much larger than that normally obtained with a silver chloride electrode of the second kind (see Chapter 2). The authors also noted improved performances, as compared with the pure silver halide pellets, of silver bromide/silver sulfide and silver iodide/silver sulfide pellets: in these cases the principal advantage of incorporating the silver sulfide is the greater strength of the pellets.

Lifetime

Most of the electrodes based on inorganic salts will give good performance for two or three years if used carefully and periodically polished. The major exception is the cyanide electrode, which, because the response mechanism involves dissolution of the membrane material, has a lifetime dependent on the strength of the cyanide solutions in which it is used: in normal use, however, it will last for several months.

If halide electrodes are used in strong (>0.1 M) halide solutions, dissolution of the membrane, due to formation of polyhaloargentate(I) complexes, also shortens the lifetime of these electrodes. Sorrentino and Rechnitz[25] showed that an iodide electrode with an AgI/Ag$_2$S membrane suffered severe attack by iodide solutions stronger than 0.1 M. An electrode used in 0.1 M sodium iodide solution was still usable after 17 h, although the membrane surface was pitted, but after the same period in 0.6 M sodium iodide solution the response had failed. Examination of a photomicrograph of the surface showed that substantial dissolution of the membrane had occurred. Johansson and Edström[42] give similar photographs of the surface of a copper electrode after attack. In such cases in which the electrode surface has been eroded the response may usually be restored by fine polishing of the membrane.

The electrodes usually fail finally because the seal around the edge of the membrane breaks or dissolves; this allows leakage to occur and, in effect, the membrane is short-circuited. Failure may result if the membrane receives a sharp knock or if the electrode is used continuously in an acid medium which dissolves the sealant (e.g. strong HF). Occasionally breakage of the membrane is the cause of failure.

APPLICATIONS

Fluoride electrode

More applications have been found for the fluoride electrode than for any other ion-selective electrode. Several hundred applications are reported, including the determination of fluoride in such varied samples as glass, teeth, soils, nuclear fuel, stack gases, urine, plant tissue and industrial effluents. For an up-to-date survey of the most recent applications either the biennial review in *Analytical Chemistry* may be consulted or the *Analytical Methods Guide* published by Orion Research Inc.

All the applications involve roughly similar sample pretreatment when the concentration of fluoride is to be measured. The sample is dissolved in an appropriate reagent, the pH is adjusted to the optimum working range for the electrode (pH 5–6) and an ionic strength adjuster is added. If any metal ions such as aluminium, iron, beryllium, zirconium, thorium or uranium, which complex fluoride, are present, a complexant is also added to remove them. These general considerations led to the development of the reagent known as T.I.S.A.B. for use in treating potable water samples; this reagent may also be used in many other applications. The current form[55] of this buffer is prepared as follows:

> Acetic acid (glacial), 57 ml
> Sodium chloride, 58 g
> CDTA, 1 g
> Dissolve and mix these constituents in about 500 ml distilled water and add strong NaOH solution (about 5 M), stirring continuously, until the pH reaches 5.5 ± 0.3. Dilute to 1 litre.

In certain samples containing high concentrations of aluminium and ferric ions it may be necessary to increase the CDTA concentration to 5 g l^{-1}. Conversely, in many British potable waters, in which the concentrations of interfering ions are low, a CDTA concentration of only 0.1 g l^{-1} has been found adequate, which allows a significant cost saving to be made on this expensive reagent. T.I.S.A.B. is added to samples at a volume ratio of 1:1. Sekerka and Lechner[50] modified T.I.S.A.B. for analyses of natural water samples containing fluoride in the range 10^{-5}–10^{-7} M because their results suggested that trace impurities of fluoride in the buffer were interfering and causing an apparently high detection limit; they eliminated the sodium chloride, as this was thought to contain most of the fluoride impurity, and used the acetate buffer to fix the ionic strength. The overall ionic strength of the buffer and thence the ratio of sample to buffer were increased to reduce interference from fluoride in the distilled water. Thus, for analysis of samples containing less than 10^{-6} M fluoride, care must be taken to use fluoride-free reagents and water.

The most important and widespread application of the fluoride electrode is in the analysis of water of all types, natural, waste, potable, sea, etc. The electrode is sufficiently well behaved to be used in automatic monitors and several such systems have been developed, primarily for water analysis. Diggens and Collis[56] used an on-line industrial analyser incorporating a fluoride electrode to measure fluoride in raw water in the range 0.01–0.15 mg l^{-1}; a fluoridated water supply was also monitored with the instrument range set to 0.1–1.5 mg l^{-1} and the monitor output could be used to operate a fluoride dosing pump. As previously described, both Erdmann[49] and Sekerka and Lechner[50] incorporated the electrode into automatic analysis systems for the analysis of natural waters. Erdmann obtained more reproducible results by spiking his T.I.S.A.B. with 0.2 mg $F^- l^{-1}$; at a standard concentration of 1.5 mg $F^- l^{-1}$ the standard deviation of the results from six replicate samples was 0.012 mg $F^- l^{-1}$ and the mean 1.492 mg $F^- l^{-1}$. Sekerka and Lechner obtained a relative standard deviation of 4.1% for ten determinations of 10^{-7} M fluoride.

The measurement of fluoride in stack gases, with an automatic analyser containing a fluoride electrode, has been described by Mascini and Liberti.[57] They absorbed the fluoride from the gas in a slow flow of citrate buffer and were able to measure concentrations of fluoride in the range 5–5000 $\mu g\ m^{-3}$; the response time of 40 min is rather long, but can be reduced if a higher response range and reagent consumption rate can be tolerated.

The determination of fluorine in organic material with a fluoride electrode after an initial combustion step has been described by Light,[58] but the method is unsatisfactory in the presence of boron. Schreiber and Frei[59] have modified the combustion step to eliminate the boron interference.

Several workers have used the fluoride electrode for the analysis of etching solutions or pickling-bath liquors. Entwistle, Weedon and Hayes[60] analysed pickling-bath liquors containing both hydrofluoric acid (5–40 g l^{-1}) and nitric acid (50–250 g l^{-1}) together with 0–20 g l^{-1} fluoride complexants. Their cell consisted of a quinhydrone electrode for sensing hydrogen ions, the fluoride electrode and a reference electrode; they showed that the potential between the hydrogen ion-sensing electrode and the fluoride electrode was controlled by the undissociated hydrofluoric acid concentration rather than the dissociated fluoride ion concentration. By reference to suitable standards they were able to determine the concentrations of both acids with a standard deviation of 1%. Johansson[61] has pointed out that the quinhydrone electrode cannot be used in samples containing iron(III) or other oxidants to which the electrode gives a redox response.

The stability constants of a range of fluoride complexes have been determined with the fluoride electrode. In one of the first papers in this field Srinivasan and Rechnitz[43] reported the results of a study of the hydrogen fluoride–fluoride ion system; they showed that the electrode response in strongly acid solutions should be interpreted solely as a response to fluoride ions and that the response to the ion HF_2^- is negligible. Moreover, on this basis there was good agreement between the stability constants quoted in the literature and those calculated from their results. Later Baumann[62] made a thorough study of the thorium–fluoride system at three temperatures. Studies have been made of many other systems and are abstracted in reviews.

The use of a lanthanum fluoride membrane for the electrochemical generation of fluoride ions with an efficiency of $99.2 \pm 0.5\%$ has been described by Durst and Ross.[21] Thus, the fluoride electrode could be used for coulometrically generating fluoride ions.

The fluoride electrode has found application as sensor in several potentiometric titrations. Relatively high concentrations of fluoride may be analysed with a precision and accuracy of about $\pm 0.1\%$, much better than is achievable by direct potentiometry, if the samples are titrated with lanthanum, either by use of a standard lanthanum nitrate solution[63] or by coulometrically generating the lanthanum,[64] and with the fluoride electrode as the end-point detector. Thorium solutions may also be used as titrant. Phosphate ions may be determined by addition of excess lanthanum nitrate and potentiometric titration of the excess with a standard fluoride solution.[65] The various problems associated with the use of ion-selective electrodes in potentiometric titrations will be discussed in Chapter 9.

Chloride, bromide and iodide electrodes

The applications of these electrodes are mostly quite straightforward in principle. The working range for the pH of samples is very large and thus the only pre-treatment which is usually necessary in determinations of sample concentration is ionic strength adjustment. The applications of the electrodes described in this section are determinations of halides; the use of the iodide electrode for sensing cyanide and mercury ions will be discussed in subsequent sections.

One of the most important applications of chloride electrodes is in the analysis of chloride in boiler water. The analysis may be made either with an ion-selective electrode or, more usually, with an electrode of the second kind, as these are cheaper and there are no redox systems in the samples to interfere. The chloride concentration to be measured is very close to the limit of detection of the silver chloride-based electrodes and, hence, the announcement of the more sensitive electrode[22] based on mercurous chloride/mercuric sulfide (Hg_2Cl_2/HgS) offers considerable promise for the development of a more satisfactory method. Bardin[66] showed that at chloride concentrations below the Nernstian limit a silver/silver chloride electrode gives a potential varying linearly with concentration, and that by means of a differential technique the electrode could be used to analyse chloride over at least a decade of concentration up to 0.3 mg l^{-1}. Such measurements are extremely difficult to make, as the electrode sensitivity is very poor. Torrance[67] buffered alkaline boiler water samples to pH 4.7 with an acetate buffer and obtained satisfactory results in the range 0.1–10 mg Cl^- l^{-1}. The standard deviation of determinations ranged from 0.04 mg l^{-1} at a chloride concentration of 0.1 mg l^{-1} to 0.23 mg l^{-1} at 10 mg l^{-1}. There was a positive bias on the results of $+0.20$ mg l^{-1} at a chloride concentration of 5.94 mg l^{-1}; this could be explained by the approximations inherent in the calculations. Both Bardin and Torrance used mercury/mercurous sulfate reference electrodes to minimize interference by the bridge solution of the reference electrode.

The determination of chloride in natural steam, tap, boiler, lake and waste waters with the Hg_2Cl_2/HgS electrode has been reported by Sekerka, Lechner and Wales.[68] The analyses of the more dilute samples were far simpler to perform and much more precise than with the silver chloride-based electrodes because of the much greater sensitivity of the electrode. Good agreement was obtained between the results of analyses of samples with the electrode and with a conventional thiocyanate colorimetric method. The authors propose a manual and an automated procedure, both suitable for measurement of chloride ions in the concentration range 5×10^{-7}–1×10^{-1} M (0.02 3500 mg l^{-1}). These authors used a double junction reference electrode with a bridge solution of 10% potassium nitrate.

A typical application of the chloride electrode is the analysis of halogenated pharmaceutical compounds described by Dessouky, Tóth and Pungor;[69] the chloride in these compounds was liberated either by simply dissolving the compound in water or by the Schöniger combustion method. The chloride content of the resultant solution was then determined either by direct potentiometry or, more accurately and precisely, by potentiometric titration with silver nitrate solution, the electrode being used to detect the end-point. De Clerq, Mertens and Massart[70] have reported the analysis of chloride in milk, an analysis which is important for the detection of mastitic or abnormal milk. Because of interference with the electrode response from constituents of the milk, principally casein, sample pretreatment is necessary with a reagent such as nitric acid or sodium tungstate and sulfuric acid to remove interferents and fix the ionic strength. Direct potentiometric determination of the chloride typically gave a relative standard deviation of about 2%, which was sufficiently good for routine analysis.

A medical application of the chloride electrode is the measurement of chloride in sweat as a screening procedure for cystic fibrosis. The measurement is conveniently carried out with a flat-ended combination chloride electrode; Orion Research Inc. have developed such an electrode specially for this purpose. Sweat is produced by electrically stimulating the sweat glands and the electrode is then placed straight on to the skin for direct measurement of the chloride concentration in the sweat. Typical results are presented by Kopito and Shwachman.[71]

The analysis of chloride in soil suspensions using an electrode of the second kind has been described by McLeod et al.;[72] a chloride-selective electrode would have to be used if appreciable concentrations of redox reagents were present in the soils. As expected, in this application use of a mercury/mercurous sulfate reference electrode with a potassium sulfate bridge solution is not satisfactory because of suspension effects causing large junction potentials; however, a calomel electrode with a potassium chloride bridge solution and a low flow-rate junction was found to give insignificant interference in the time needed to take the measurements, because of the relatively high chloride level in the samples. The authors developed a system for the simultaneous measurement of pH, chloride content and conductivity of the soil suspensions, and obtained a coefficient of variation in the chloride measurement, including sampling errors, of 7.27%.

Applications of the chloride electrode and, to a lesser extent, the bromide electrode are frequently limited by interference by other halides and pseudo-halides in samples; it is particularly important in the development of a method using these electrodes that the concentrations of interferents such as iodide and sulfide ions be sufficiently low. Such interference with the chloride electrode may often be removed by oxidation of the interferents with excess permanganate and removal of the excess with hydrogen peroxide[73] or by oxidation with chromic acid.[74] Sulfide interference may also be prevented by acidification to convert sulfide ions, S^{2-}, to hydrogen sulfide, H_2S. If it is not possible to remove the interference, use of the chloride electrode based on an organic ion exchanger may be appropriate (see Chapter 6).

Like the fluoride electrode, other halide electrodes have been used for the determination of stability constants of halide complexes by several workers. A recent example is the study by Duff,[75] using chloride and bromide electrodes, of benzene–1,2-diamine complexes with nickel halides.

Ion-selective electrodes may also be used to follow chemical reactions; a particularly interesting example of this is the work of Noyes, Field and Kőrös[76] and Kőrös and Burger[77] on Zhabotinsky-type oscillating reactions. A typical reaction involves the catalysed oxidation of malonic acid with bromate ions in a sulfuric acid medium; typical catalysts are cerium(III), manganese(II) or ferroin. As the reaction proceeds, bromide ions are formed which inhibit the oxidation of the malonic acid but are in turn consumed by reaction with the bromate ions; the result of a complex set of reactions is that the bromide ion concentration, and also the redox potential of the system, oscillates at a frequency of about one cycle in two minutes (with appropriate reagent concentrations) as shown in Fig. 5.9. A bromide electrode and a redox electrode may therefore be used to follow the reaction for mechanistic studies.

Analytical applications of electrodes for bromide and iodide measurement are relatively few and no special experimental conditions apply. One ingenious application of the iodide electrode is in the determination of trace metals by measuring the catalysed rate of reaction between hydrogen peroxide and iodide ions. Altinata and Pekin[78] determined molybdenum and tungsten in this way.

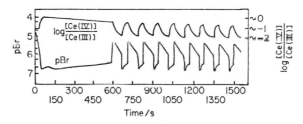

Fig. 5.9 Potentiometric traces showing the responses of a redox electrode and a bromide electrode used to follow an oscillating chemical reaction.[76] Initial concentrations: $[CH_2(COOH)_2] = 0.032$ M; $[KBrO_3] = 0.063$ M; $[KBr] = 1.5 \times 10^{-5}$ M; $[Ce(NH_4)_2(NO_3)_5] = 0.001$ M; $[H_2SO_4] = 0.8$ M

The behaviour in partially non-aqueous solvents of halide electrodes with silicone rubber matrices has been investigated by Kazarjan and Pungor[79] and could be explained largely in terms of the varying solubilities of the silver compounds in the different media. Elbakai *et al.*[80] also studied the behaviour of halide electrodes in methanolic and ethanolic solutions: they commented on the substantial uncertainty of liquid junction potentials in these media and demonstrated changes in this potential of more than 30 mV as the solution composition changed.

Silver/sulfide electrodes

Samples in which sulfide is to be determined must be made alkaline to convert all H_2S and HS^- species to sulfide ions, S^{2-}; it is also advisable to add a reductant such as ascorbic acid to prevent oxidation of the sulfide by air. Orion Research Inc.[81] recommend a buffer known as S.A.O.B. (Sulphide Anti-Oxidant Buffer) for sample pretreatment. This is added to the sample at a volume ratio of 1:1. This buffer is made from sodium hydroxide (40 g l^{-1}), sodium salicylate (160 g l^{-1}) and ascorbic acid (36 g l^{-1}). Baumann[82] found that the usefulness of this buffer was limited to the treatment of samples containing more than 0.1 μg S^{2-} l^{-1}, because of impurities of copper and chromium in the sodium salicylate which complexed the sulfide. Baumann also developed a preconcentration method for extending the range of measurement. Samples containing less than 30 μg S^{2-} l^{-1} (10^{-6} M) were treated with a reagent containing zinc acetate and sodium carbonate, which coprecipitated zinc sulfide with zinc hydroxide. The precipitate was separated and redissolved in a small quantity of alkaline EDTA plus ascorbic acid. After a tenfold preconcentration by this method, a sulfide sample originally containing 1.8 μg S^{2-} l^{-1} (6×10^{-7} M) could be measured with a relative standard deviation of 6%.

Sulfide has been measured in many types of industrial waters by use of an ion-selective electrode. An early example was the monitoring of the oxygen purging of sulfide in black liquor solutions reported by Light;[83] it is interesting to note that, as the sulfide concentration decreased below about 10^{-5} M, a sharp increase in potential was observed consistent with the behaviour shown in Fig. 5.3. From the point of view of method development, these and other applications present few problems because of the high selectivity of the electrode (only mercury ions interfere).

Mor et al.[84] used a sulfide electrode to study the sulfide ions in sea-water. They initially determined the stability constants of the hydrogen ion–sulfide ion complexes in sea-water. The sulfide ion concentration at several pH values was then calculated and a calibration graph was plotted. The resultant line was compared with the similar plot obtained from measurement of the sulfide ion activity in phosphate buffers at different pH values. Hence, the activity coefficient of sulfide ions in sea-water was calculated to be about 0.2.

A silver/sulfide electrode has been used by Alexander and Rechnitz[85] in the automated determination of serum protein using a Technicon AutoAnalyzer system. The method involved the addition of excess silver ions to a protein sample undergoing denaturation; the silver ions react with the disulfide content of the protein and the excess silver ion concentration is measured after a fixed time.

Pápay et al. have used sulfide electrodes as indicator electrodes in the potentiometric titration of thiourea[86] and thioacetamide[87] with standard silver nitrate solutions. The titrations were carried out in strong alkaline solutions and the products of the titrations were analysed. In general, this silver/sulfide electrode is the most suitable electrode for potentiometric titrations with silver nitrate.

Applications involving the use of the electrode for determining silver in samples are straightforward in principle, as the electrode is very tolerant to pH and highly selective for silver ions. Durst and Duhart[88] used the electrode for studying the rate of adsorption of silver ions on to selected surfaces; after 30 days the amount of silver ions adsorbed increased in the order Vycor < polythene < PTFE < silicone-coated Pyrex < Pyrex, although the adsorption of silver ions on to the PTFE had an induction period of two days. Müller, West and Müller,[89] measuring in the range 10^{-6}–10^{-7} M Ag$^+$, showed that silver ions could be determined with an accuracy of 0.2% by use of a procedure akin to multiple standard addition (see Chapter 9).

Single junction silver/silver chloride reference electrodes are unsuitable for use with the silver/sulfide electrode because of the leakage of silver species through the liquid junction; this can result in blockage of the junction with silver sulfide in sulfide samples or contamination with excess silver of silver samples. As an alternative, a single junction calomel reference electrode is suitable.

Cyanide electrode (AgI-based)

The major problem with the cyanide electrode is that the membrane dissolves away and the rate of dissolution is rapid in cyanide solutions more concentrated than 10^{-3} M. Even in the more dilute solutions relatively frequent polishing of the membrane is necessary to expose a fresh surface of silver iodide and maintain optimum electrode performance. To circumvent this problem it is now conventional in automatic analysers to replace the silver iodide electrode with a silver/sulfide electrode and rely on the sensing of silver in equilibrium with $Ag(CN)_2^-$ on the electrode surface. The $Ag(CN)_2^-$ ions may be either added to the sample[90] or formed on the membrane surface by attack of the cyanide ions on the silver defect ions;[13] in either case the system gives a response slope of 120 mV decade^{-1} change in cyanide ion activity instead of 60 mV decade^{-1} obtained from the silver iodide electrode. The direct response of the silver/sulfide electrode to cyanide ions, with addition of $Ag(CN)_2^-$ to the sample, appears to work much more satisfactorily in the controlled environment of an automatic analyser than in manual methods.

For manual laboratory determinations of cyanide the iodide electrode is still used. Samples must be made alkaline (pH > 11) to convert any HCN present to CN^-. It has been shown by Tóth and Pungor[91] that metal cyanide complexes more stable than $Ag(CN)_2^-$ (e.g. $Ni(CN)_4^{2-}$) are not sensed by the electrode, but the electrode does sense part of the cyanide in less stable cyanides (e.g. $Zn(CN)_4^{2-}$ or $Cd(CN)_4^{2-}$). A quantitative treatment is given by Mascini and Napoli.[92] Sample pretreatment is therefore necessary for measurement of total cyanide concentration. This is easily possible in solutions containing unstable and labile cyanide complexes, since addition of EDTA usually suffices to decomplex the cyanide. However, if samples contain stable complexes such as $Fe(CN)_6^{4-}$ or $Ni(CN)_4^{2-}$, prolonged heating with complexant, treatment with ultraviolet radiation, etc. is required to decompose the complexes.

Applications of the cyanide electrode include the analysis of plating-bath solutions, plating wastes, fruit brandies and sewage. Also, the electrode may be used to analyse plant material (e.g. Sudan grass, almond, sorghum) for cyanogenic glucosides; the procedure involves hydrolysis of the material followed by acidic distillation of the cyanide into an alkali and measurement with the electrode. Blaedel et al.[93] found it necessary to use a standard addition technique to get accurate results (about 2% relative error).

Fleet and von Storp[94] report that the potential of the cyanide electrode is quite strongly dependent on the flow rate of sample across the membrane; hence, care must be taken to keep the stirring rate constant between standards and samples. The authors also studied the performance of the electrode in a modified Technicon AutoAnalyzer system. High flow rates were required to achieve fast electrode response. At a flow rate of 10 ml min^{-1} the response time for the step 10^{-4} M to 10^{-3} M CN$^-$ was 30 s and 75 s for the converse change; however, if the flow rate was increased to 40 ml min^{-1}, the response times were 6 s and 20 s, respectively.

Mercury electrode (AgI-based)

The silver iodide-based electrodes may be used to determine mercury or, more importantly, to serve as indicator electrode in mercurimetric titrations. The main limitation with the use of the electrode in potentiometric titrations is the interference caused by chloride ions, which should not be present in samples in more than trace amounts. Chloride ions may be removed from samples to be titrated by an ion-exchange process.[41] It is also important to ensure that the pH of samples in which mercury is to be determined directly is kept in the range 4–5.

The electrode has been used as indicator electrode in the titration of soluble thiol groups in flour with o-hydroxymercuribenzoate.[95] The analysis may be used to determine the degree of cooking of wheat germ used in bread making. A mercury/mercurous sulfate reference electrode with a 1 M sodium sulfate bridge solution was used to prevent chloride interference.

Copper electrode

The copper electrode is the most successful of the heavy metal electrodes and several applications have been developed. In general, the main considerations in developing a method using the electrode are that the sample should have the correct pH (in the range 3–7) and oxidants such as Fe(III) should be absent; addition of a mild reductant to the buffering reagent helps to prolong the useful life of the membrane.

An example of careful work with the copper electrode is the determination of copper by a standard addition technique down to 9 μg l^{-1} (1.5×10^{-7} M) in tap-water and natural waters reported by Smith and Manahan.[48] They found that the sensitivity of the determination was limited by the concentration of copper in their purified water and reagents. A complexing anti-oxidant buffer was developed, which was added to samples in a volume ratio of 1:1; the composition of the buffer was

> 0.05 M sodium acetate
> 0.05 M acetic acid
> 0.02 M sodium fluoride
> 2×10^{-3} M formaldehyde

Ultra-pure reagents were used in the preparation of the buffer but even so a background level of 1.2 μg Cu^{2+} l^{-1} was found to be typical. The acetate buffer fixed the pH and ionic strength of samples and also complexed a fixed proportion of the copper ions. The complexation of the copper by acetate served to decomplex the copper from all but very strong complexants in the sample and minimized adsorption of copper on to the walls of the sample container. For analysis by standard addition techniques it is relatively unimportant what fraction of the copper ions are free but crucial that that fraction remain constant throughout the experiment. The fluoride ions complex any ferric ions present and the formaldehyde provides the reducing medium. At the lower end of the range of the method a spike of 9.0 μg Cu^{2+} l^{-1} was added to natural water samples containing 3.3–46.8 μg Cu^{2+} l^{-1}; the average recovery was 102.9%, with a standard deviation of 7.5%.

The response of the electrode in samples of extreme dilution (down to 10^{-9} M) has been studied by Blaedel and Dinwiddie;[47,96] they report the procedures necessary to prepare the reagents and the electrode for use at such levels.

With less difficulty copper can be measured in waste waters, plating liquors, soil and etching baths. The electrode has also proved very useful as indicator electrode in the compleximetric titration of ions such as Ca^{2+}, Zn^{2+}, Ni^{2+} [31,97] and Cu^{2+} [31,98] with EDTA, NTA, etc. Šucha and Suchánek[99] determined aluminium by adding excess CDTA and back-titrating the excess with a copper solution using a copper electrode to detect the end-point. Stability constants for the Cu(II)/ethane-1-hydroxy-1,1-diphosphoric acid system have been determined by use of a copper electrode.[100]

Cadmium and lead electrodes

There are few reported applications of the cadmium electrode, probably because of the unsatisfactory performance of the electrode and the relative ease of measuring cadmium by other analytical techniques. The electrode is probably most useful as indicator electrode in the potentiometric titration of large concentrations of cadmium with such titrants as EDTA or NTA.[33,38]

The lead electrode has also been used primarily in potentiometric titrations, and, in particular, has been used for the titration of sulfate with standard lead perchlorate solution. A procedure for the titration of sulfate in natural waters has been developed.[101] The titration has to be carried out in a partially nonaqueous solvent such as methanol or ethanol in order to reduce the solubility of the lead sulfate and, hence, get a sharper end-point. The choice of nonaqueous solvent has been discussed by Selig and Salomon;[102] they preferred methanol or ethanol to p-dioxane, which has previously been recommended, on the grounds that p-dioxane generates peroxides on standing and these poison the electrode. Ross and Frant[103] were able to determine the end-point in titrations in p-dioxane to $\pm 0.2\%$ for sulfate samples stronger than 5×10^{-4} M and to $\pm 1\%$ at 10^{-5} M. Excess chloride ions or nitrate ions in samples produced an error, as did the presence of copper, mercury or silver ions.

Sulfur may also be measured after conversion to sulfate. Thus, Heistand and Blake[104] determined sulfur in petroleum by pyrolysing the sample and titrating the sulfate formed; a curious feature of their work is that they polarized the electrode with a constant current of 0.5 μA in order to get a sharper end-point.

Other ions which may be potentiometrically titrated with a standard lead solution include $C_2O_4^{2-}$, $Fe(CN)_6^{4-}$, WO_4^{2-} and $P_2O_7^{4-}$,[34] but care must be taken to ensure that no oxidants are present in the samples, as these destroy the electrode response. In those applications where the electrodes are used in partially nonaqueous media it is most satisfactory to use a double junction reference electrode with a bridge solution of composition closely similar to that of the titrand.

In the preparation of a calibration graph for the lead electrode the standard solutions should be made from lead perchlorate, as the perchlorate anions do not appreciably complex the lead ions. Lead nitrate may also be used for solutions more dilute than 10^{-3} M, but at higher concentrations errors result from the association of the lead and nitrate ions.

No important applications of the thiocyanate electrode have been reported.

REFERENCES

1. E. Pungor and K. Tóth, *Analyst* **95**, 625 (1970).
2. M. S. Frant, U.S. Patent No. 3,431,182 (4th March 1969).
3. M. S. Frant and J. W. Ross, *Science, N.Y.* **154**, 3756 (1966).
4. M. S. Mohan and G. A. Rechnitz, *Anal. Chem.* **45**, 1323 (1973).
5. M. J. D. Brand and G. A. Rechnitz, *Anal. Chem.* **42**, 478 (1970).
6. E. Pungor, *Anal. Chem.* **39** (13), 28A (1967).
7. H. M. Stahr and D. O. Clardy, *Anal. Lett.* **6**, 211 (1973).
8. A. Marton and E. Pungor, *Anal. Chim. Acta* **54**, 209 (1971).
9. R. P. Buck and V. R. Shepard, *Anal. Chem.* **46**, 2097 (1974).
10. C. Wagner, *Z. Elektrochem.* **63**, 1027 (1959).
11. J. Růžička, C. G. Lamm and J. Tjell, *Anal. Chim. Acta* **62**, 15 (1972).
12. M. Koebel, *Anal. Chem.* **46**, 1559 (1974).
13. W. E. Morf, G. Kahr and W. Simon, *Anal. Chem.* **46**, 1538 (1974).
14. R. A. Durst, in *Ion-Selective Electrodes*, R. A. Durst, editor, Chapter 11. N.B.S. Spec. Publ. No. 314, Washington, D.C. (1969).
15. L. G. Sillén and A. E. Martell, *Stability Constants*, Spec. Publ. No. 17, The Chemical Society, London (1964).
16. Orion Research Inc., *Newsletter* **3**, 25 (1971).
17. J. W. Ross, in *Ion-Selective Electrodes*, R. A. Durst, editor, Chapter 2. N.B.S. Spec. Publ. No. 314, Washington, D.C. (1969).
18. D. J. Crombie, G. J. Moody and J. D. R. Thomas, *Talanta* **21**, 1094 (1974).
19. K. J. Vetter, *Electrochemical Kinetics*, trans. ed. S. Bruckenstein and B. Howard. Academic Press, New York (1967).
20. J. Buffle, N. Parthasarathy and W. Haerdi, *Anal. Chim. Acta* **68**, 253 (1974).
21. R. A. Durst and J. W. Ross, *Anal. Chem.* **40**, 1343 (1968).
22. J. F. Lechner and I. Sekerka, *J. Electroanal. Chem.* **57**, 317 (1974).
23. I. Sekerka and J. F. Lechner, *J. Electroanal. Chem.* **69**, 339 (1976).
24. J. Bagg and G. A. Rechnitz, *Anal. Chem.* **45**, 271 (1973).
25. M. H. Sorrentino and G. A. Rechnitz, *Anal. Chem.* **46**, 943 (1974).
26. K. Tóth and E. Pungor, *Anal. Chim. Acta* **51**, 221 (1970).
27. T.-M. Hseu and G. A. Rechnitz, *Anal. Chem.* **40**, 1054 (1968).
28. C. L. Jamson, unpublished work in the E.I.L. laboratory.
29. E. Baumann, *Anal. Chim. Acta* **54**, 189 (1971).
30. R. Blum and H. M. Fog, *J. Electroanal. Chem.* **34**, 485 (1972).

31. E. H. Hansen, C. G. Lamm and J. Růžička, *Anal. Chim. Acta* **59**, 403 (1972).
32. J. Pick, K. Tóth and E. Pungor, *Anal. Chim. Acta* **61**, 169 (1972).
33. J. Růžička and E. H. Hansen, *Anal. Chim. Acta* **63**, 115 (1973).
34. E. H. Hansen and J. Růžička, *Anal. Chim. Acta* **72**, 365 (1974).
35. J. N. Butler, in *Ion-Selective Electrodes*, R. A. Durst, editor, Chapter 5. N.B.S. Spec. Publ. No. 314, Washington, D.C. (1969).
36. P. A. Evans, G. J. Moody and J. D. R. Thomas, *Lab. Pract.* **20**, 644 (1971).
37. J. Pick, K. Tóth and E. Pungor, *Anal. Chim. Acta* **65**, 240 (1973).
38. M. Mascini and A. Liberti, *Anal. Chim. Acta* **64**, 63 (1973).
39. M. J. D. Brand, J. J. Militello and G. A. Rechnitz, *Anal. Lett.* **2**, 523 (1969).
40. M. Mascini and A. Liberti, *Anal. Chim. Acta* **60**, 405 (1972).
41. Orion Research Inc., *Newsletter* **2**, 41 (1970).
42. G. Johansson and K. Edström, *Talanta* **19**, 1623 (1972).
43. K. Srinivasan and G. A. Rechnitz, *Anal. Chem.* **40**, 509 (1968).
44. K. Tóth, I. Gavallér and E. Pungor, *Anal. Chim. Acta* **57**, 131 (1971).
45. K. Tóth and E. Pungor, *Anal. Chim. Acta* **64**, 417 (1973).
46. R. Rangarajan and G. A. Rechnitz, *Anal. Chem.* **47**, 324 (1975).
47. W. J. Blaedel and D. E. Dinwiddie, *Anal. Chem.* **46**, 873 (1974).
48. M. J. Smith and S. E. Manahan, *Anal. Chem.* **45**, 836 (1973).
49. D. E. Erdmann, *Env. Sci. Technol.* **9**, 252 (1975).
50. I. Sekerka and J. F. Lechner, *Talanta* **20**, 1167 (1973).
51. L. E. Negus and T. S. Light, *Instrum. Technol.* 23 (December 1972).
52. E. Lindner, K. Tóth and E. Pungor, in *Ion-Selective Electrodes*, E. Pungor, editor, p. 205. Akadémiai Kiadó, Budapest (1973).
53. A. J. de Bethune, T. S. Light and N. Swendeman, *J. Electrochem. Soc.* **106**, 616 (1959).
54. J. W. Ross, M. S. Frant and J. H. Riseman, U.S. Patent No. 3,563,874 (16th February, 1971).
55. J. E. Harwood, *Water Res.* **3**, 273 (1969).
56. A. A. Diggens and D. E. Collis, *Water Treat. Exam.* **18**, 192 (1969).
57. M. Mascini and A. Liberti, *Gazz. Chim. Ital.* **103**, 989 (1973).
58. T. S. Light, *Anal. Chem.* **41**, 107 (1969).
59. B. Schreiber and R. W. Frei, *Mikrochim. Acta* 219 (1975).
60. J. R. Entwistle, C. J. Weedon and T. J. Hayes, *Chem. Ind.* (*London*) 433 (1973).
61. G. Johansson, *Anal. Chim. Acta* **77**, 283 (1975).
62. E. Baumann, *J. Inorg. Nucl. Chem.* **32**, 3823 (1970).
63. J. J. Lingane, *Anal. Chem.* **39**, 881 (1967).
64. D. J. Curran and K. S. Fletcher, *Anal. Chem.* **41**, 267 (1969).
65. Orion Research Inc., *Newsletter* **3**, 2 (1971).
66. A. A. Bardin, *Zavod. Lab.* **28**, 967 (1962).
67. K. Torrance, *Analyst* **99**, 203 (1974).
68. I. Sekerka, J. F. Lechner and R. Wales, *Water Res.* **9**, 663 (1975).
69. Y. M. Dessouky, K. Tóth and E. Pungor, *Analyst* **95**, 1027 (1970).
70. H. L. De Clerq, J. Mertens and D. L. Massart, *J. Agric. Food Chem.* **22**, 153 (1974).
71. L. Kopito and H. Shwachman, *Pediatrics* **43**, 794 (1969).
72. S. McLeod, H. C. T. Stace, B. M. Tucker and P. Bakker, *Analyst* **99**, 193 (1974).
73. J. Havas, E. Papp and E. Pungor, *Acta Chim. Hung.* **58**, 9 (1968).
74. J. C. Van Loon, *Analyst* **93**, 788 (1968).
75. E. J. Duff, *J. Inorg. Nucl. Chem.* **36**, 1167 (1974).
76. R. M. Noyes, R. J. Field and E. Kőrös, *J. Am. Chem. Soc.* **95**, 1394 (1972).
77. E. Kőrös and M. Burger, in *Ion-Selective Electrodes*, E. Pungor, editor, p. 191. Akadémiai Kiado, Budapest (1973).
78. A. Altinata and B. Pekin, *Anal. Lett.* **6**, 667 (1973).

79. N. A. Kazarjan and E. Pungor, *Anal. Chim. Acta* **51**, 213 (1970).
80. A. M. Elbakai, G. J. Kakabadse, M. N. Khayat and D. Tyas, *Proc. Anal. Div. Chem. Soc.* **12**, 83 (1975).
81. Orion Research Inc., Applications Bulletin No. 12, 1969.
82. E. Baumann, *Anal. Chem.* **46**, 1345 (1974).
83. T. S. Light, in *Ion-Selective Electrodes*, R. A. Durst, editor, Chapter 10. N.B.S. Spec. Publ. No. 314, Washington, D.C. (1969).
84. E. Mor, V. Scotto, G. Marcenaro and G. Alabiso, *Anal. Chim. Acta* **75**, 159 (1975).
85. P. W. Alexander and G. A. Rechnitz, *Anal. Chem.* **46**, 860 (1974).
86. M. K. Pápay, K. Tóth and E. Pungor, *Anal. Chim. Acta* **56**, 291 (1971).
87. M. K. Pápay, K. Tóth, V. Izvekov and E. Pungor, *Anal. Chim. Acta* **64**, 409 (1973).
88. R. A. Durst and B. T. Duhart, *Anal. Chem.* **42**, 1002 (1970).
89. D. C. Müller, P. W. West and R. H. Müller, *Anal. Chem.* **41**, 2038 (1969).
90. M. S. Frant, J. W. Ross and J. H. Riseman, *Anal. Chem.* **44**, 2227 (1972).
91. K. Tóth and E. Pungor, *Anal. Chim. Acta* **51**, 221 (1970).
92. M. Mascini and A. Napoli, *Anal. Chem.* **46**, 447 (1974).
93. W. J. Blaedel, D. B. Easty, A. Laurens and T. R. Farrell, *Anal. Chem.* **43**, 890 (1971).
94. B. Fleet and H. von Storp, *Anal. Chem.* **43**, 1575 (1971).
95. K. M. Bailey, unpublished work (Weston Research Laboratories Ltd).
96. W. J. Blaedel and D. E. Dinwiddie, *Anal. Chem.* **47**, 1070 (1975).
97. J. W. Ross and M. S. Frant, *Anal. Chem.* **41**, 1900 (1969).
98. E. W. Baumann and R. M. Wallace, *Anal. Chem.* **41**, 2072 (1969).
99. L. Šůcha and M. Suchánek, *Anal. Lett.* **3**, 613 (1970).
100. H. Wada and Q. Fernando, *Anal. Chem.* **43**, 751 (1971).
101. Orion Research Inc., *Analytical Methods Guide*, 7th edn, p. 30 (1975).
102. W. Selig and A. Salomon, *Mikrochim. Acta* 663 (1974).
103. J. W. Ross and M. S. Frant, *Anal. Chem.* **41**, 967 (1969).
104. R. N. Heistand and C. T. Blake, *Mikrochim. Acta* 212 (1972).

CHAPTER 6

ELECTRODES BASED ON ION EXCHANGERS AND NEUTRAL CARRIERS

A wide range of ion-selective electrodes based on organic ion exchangers and neutral carriers has been developed and reported. Because of the enormous number of ion-exchange systems available, ion-selective electrodes may be developed for almost any anion or cation from chloride to dodecyl sulfate and from calcium to uranyl ions. However, not all these electrodes are equally satisfactory and, in particular, the selectivities of some are poor. The number of ions for which suitable neutral carriers have been reported is much more limited; so far electrodes have been developed only for the cations of potassium, calcium, barium, sodium and ammonium. These neutral carrier electrodes are, to generalize, more selective than the ion-exchange electrodes. Despite the number of electrodes reported, comparatively few types are available commercially and even fewer are at all widely used: the most popular are the nitrate, potassium and calcium electrodes.

The ion-exchange electrodes were developed before the neutral carrier electrodes and, because the ion exchanger was usually incorporated into the electrode in solution in an organic solvent, they were commonly known as 'liquid ion-exchange electrodes'. This name has now fallen into disuse, not only because of its length, but also because it is now fashionable to incorporate the ion exchanger in an inert matrix such as PVC, polythene or silicone rubber to obtain a 'solid' membrane. Thus, the electrodes are now commonly called simply 'ion-exchange electrodes'. This term, 'ion-exchange electrode', is taken to mean an electrode having as active material a charged ion exchanger consisting of large organic molecules: the term is not used to denote any electrode for which ion exchange at the membrane–solution interface contributes to the electrode potential, as it would then include the glass electrodes and inorganic salt-based electrodes. Either liquid or solid membrane construction may also be used for the neutral carrier electrodes. Diagrams of both constructions are shown in Fig. 6.1.

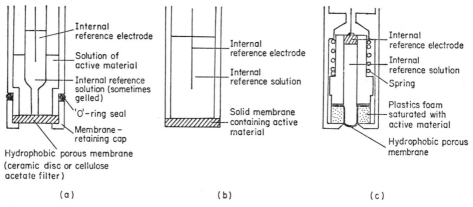

Fig. 6.1 Various constructions of the sensing tips of ion-selective electrodes based on organic ion exchangers or neutral carriers: (a) liquid membrane (similar to the Corning construction); (b) solid membrane, sealed on to the body; (c) liquid membrane (new Orion type)

The advantages and disadvantages of liquid and solid membranes have been argued at length in the literature: the proponents of the solid membranes have succeeded in filling more journal pages with these discussions but reports of applications refer almost entirely to electrodes with the liquid membranes. A resumé of the differences between the two types of ion-exchange membranes is given in Table 6.1; information in the literature is frequently conflicting on some points of performance and so a selection has had to be made, or a range given, to allow for different types of use. On the assumption that commercial electrodes are used, the time taken to make the solid membranes is not included in the comparison.

An important point made by the protagonists of the solid membranes has been the inconvenience of handling and servicing the electrodes with liquid membranes, which involves the handling of the obnoxious ion-exchange solutions, the correct positioning of tiny Millipore or Nucleopore filters, and so forth. The

TABLE 6.1
Comparison of liquid and solid membranes

	Liquid membrane	Solid membrane
Wt of active material incorporated	about 100 mg	about 1 mg
Ease of electrode assembly	dependent on design	easy
Selectivity	as solid	as liquid
Response speed	see text	
Effect of strong interferents	Prolonged drift	Short drift or failure
Typical E° drift	1 mV day^{-1}	1 mV week^{-1}
Sample contamination by electrode	Slight	Negligible
Lifetime	1–2 months	1 week–2 months

majority of these criticisms apply primarily to the obsolete Orion electrode, which had a troublesome design; the Corning and the new Orion electrodes, for example, are just as easy to service as those with solid membranes. The ingenious design of the new Orion electrodes allows a new membrane assembly to be simply screwed into the electrode body in place of the exhausted assembly and no handling of the liquid ion exchanger solution is necessary at all; the solid membranes are usually replaced in a similar way.

The most fundamental differences between the liquid and solid membranes arise from the greater mobility of the species and the larger quantity of active material present in the liquid membranes.

For the electrodes based on neutral carriers, the greater mobility of the active material and ions in the liquid membranes results in a much shorter response time (see below), although for the electrodes containing charged organic ion exchangers the solid membranes respond slightly faster.

When a liquid membrane electrode is immersed in a sample containing a large concentration of interferent, the interferent is carried right into the bulk of the membrane and takes a long time to diffuse out again when the source of interferent is removed. However, when a solid membrane is used, the interferent is usually left in the surface of the membrane and is quickly removed when the membrane is immersed in a sample free from interferent; if the membrane is exposed to the interferent sufficiently long, a significant amount of it is eventually carried into the bulk of the membrane and the performance is only recovered by replacing the membrane. Hence, the performance of the electrodes with solid membranes is usually more satisfactory in the presence of interferents because the membrane is destroyed more slowly; however, it is still necessary to remove interferents from samples before an accurate measurement of the determinand activity can be made. Thus, for direct potentiometric determinations, the samples should be pretreated so that any interferents originally present have been removed before the electrode is immersed in them. When the electrode is used as a sensor in a potentiometric titration, the situation is more complicated, as sometimes the excess of titrand present after the end-point may interfere with the response of the electrodes; in such cases it is probably better to use a solid membrane electrode and to store the electrode, between titrations, in a dilute solution of determinand.

The ion exchangers are all very slightly soluble in water. In the liquid membranes more ion exchanger can diffuse from the reservoir to the surface of the membrane to replace that which has dissolved; similarly, if the electrode surface has become contaminated, the surface can be wiped with, for example, cotton wool to produce a fresh surface. For solid membranes, when the ion exchanger has dissolved out of the surface, the membrane must be discarded. The solubility of the ion exchanger in samples depends on the concentration of determinand and also on pH. In general, the higher the concentration of determinand the greater will be the solubility: the effect of pH cannot be similarly generalized, although the more extreme the pH the higher the solubility is likely to be. The solubility will also be drastically affected if the sample solvent is changed from water to a more lipophilic solvent.

These factors affect the membrane lifetime, which is set by the total volume of samples with which the electrode can come to equilibrium before the active material is irreplaceably dissolved out of the membrane. This parameter is important for the analyst, as it allows an estimate to be made of the total number of samples which can be analysed, or the length of time the electrode may be used in a continuous analyser, before failure; the length of time an electrode will last if stored in a beaker of determinand and checked once a week or once a month is often quoted in papers but is almost completely irrelevant. A direct comparison of ion-exchange electrodes with solid (PVC) and liquid (Orion old type) membranes has been made[1] using an industrial analyser to monitor continuously calcium and nitrate in tap-water. Several commercial and university-made solid membrane electrodes were used for both determinands, but all failed completely within a week, whereas the Orion liquid membrane electrodes usually lasted between one and two months before needing attention. The solid membrane electrodes also had a shorter life when used for a series of manual determinations in the laboratory. This shorter life is a direct result of the much smaller amount of ion exchanger in the solid membrane.

The same considerations also apply to the two types of membrane containing neutral carriers, although the difference in lifetime is here not so pronounced because of the smaller solubility of the active materials in samples.

Mascini and Pallozzi[2] have pointed out that the life of solid membranes may also be limited by the volatility of the plasticizer used with the active material. If the electrodes are exposed to high temperatures or the membranes stored for several months at room temperature, evaporation of the plasticizer will cause the electrode to fail; the original performance may be restored by simply adding a drop of the plasticizer to the membrane. The life of liquid membranes is also limited by the volatility of the solvent.

To summarize, if an electrode is to be used very frequently or for continuous monitoring, the longer life of a liquid membrane makes it more satisfactory. For occasional use in direct measurements, use in potentiometric titrations or where there is a danger of the electrode being subjected to strong interference, a solid membrane is better. Frequently there will be only a marginal advantage in using one form compared with another. Either type of electrode may be simply constructed in the laboratory as described in Chapter 9, if the appropriate active material is available.

Several ion exchangers and plasticizers or solvents have been used in electrodes for the most popular determinands. A list of the more important and successful electrodes based on organic ion-exchangers together with their constitution is given in Table 6.2.

A similar list of the components of electrodes based on neutral carriers is given in Table 6.3. A sodium electrode based on a neutral carrier has also been reported,[20] but, in general, it compares unfavourably in performance with the

TABLE 6.2
Electrodes based on ion exchangers

Determinand	Form of membrane	Active material	Solvent or plasticizer	Ref.	Comments
Ca^{2+}	(a) liquid	calcium di-(n-decyl) phosphate	di-(n-octylphenyl) phosphonate	3	Orion electrode
	(b) solid (PVC)	calcium di-(n-decyl) phosphate	di-(n-octylphenyl) phosphonate	4, 5	
	(c) solid (PVC) 'Selectrode'	calcium di-(n-octylphenyl) phosphate	di-(n-octylphenyl) phosphonate	6	Improved sensitivity and pH range
NO_3^-	(a) liquid	tridodecylhexadecylammonium nitrate	n-octyl-2-nitrophenyl ether		Corning electrode
	(b) solid (PVC)	tridodecylhexadecylammonium nitrate	n-octyl-2-nitrophenyl ether	7	
	(c) liquid	tris(substituted 1,10-phenanthroline) nickel(II) nitrate	p-nitrocymene	8	Orion electrode
	(d) solid PVC	tris(substituted 1,10-phenanthroline) nickel(II) nitrate	p-nitrocymene	7	
	(e) liquid	tris(substituted 1,10-phenanthroline) nickel(II) nitrate	p-nitrocymene	9	Convenient construction for home-made electrodes
ClO_4^-	liquid	tris(substituted 1,10-phenanthroline) iron(II) perchlorate	p-nitrocymene	8	
BF_4^-	liquid	tris(substituted 1,10-phenanthroline) nickel(II) tetrafluoroborate	p-nitrocymene decanol	10	
Divalent cations ('water hardness')	liquid	calcium di-(n-decyl) phosphate	trifluoroacetyl-p-butylbenzene	11	Not commercially available
CO_3^{2-}	liquid	tri(n-octyl)methylammonium chloride	o-dichlorobenzene	12	Not commercially available
U (VI)	(a) liquid	methylene blue–uranyl tribenzoate	diamylamyl phosphonate	13	Not commercially available
	(b) solid (PVC)	di(2-ethylhexyl) phosphoric acid			
Cl^-	liquid	dimethyl-dioctadecylammonium chloride	?		

TABLE 6.3
Electrodes based on neutral carriers

Determinand	Form of membrane	Active material	Solvent or plasticizer	Ref.	Comments
K^+	(a) liquid	valinomycin	diphenyl ether	14	Philips electrode (similar to several other manufacturers' electrodes)
	(b) solid (PVC)	valinomycin	dioctyladipate	15	No solvent
	(c) solid (silicone rubber)	valinomycin	—	16	
NH_4^+	liquid	nonactin/monactin	tris(2-ethylhexyl) phosphate	17	Philips electrode
Ca^{2+}	solid (PVC)	see Ref. 18	o-nitrophenyl octyl ether	18	Not commercially available
Ba^{2+}	liquid	nonylphenoxy poly (ethylene oxy) ethanol Ba^{2+} (tetraphenylborate)$_2$	p-nitroethylbenzene	19	Not commercially available

sodium-responsive glass electrode, although the response time is said to be less affected by proteins. Cyclic polyethers (crown compounds) have been widely studied as potentially suitable neutral ligands for use in ion-selective membranes, primarily in potassium electrodes, but so far their performance has not equalled that of the antibiotic-type neutral ligands such as valinomycin (see, e.g., Ref. 21). Potassium electrodes have also been prepared from the ion exchangers potassium tetraphenylborate and potassium tetrachlorophenylborate, but both are much less selective for potassium than valinomycin; the values of k_{KNa} are larger by a factor of at least 10^2. Baumann[22] has reported on the performance of a strontium electrode; it does not appear to be sufficiently selective, particularly in the presence of monovalent cations, to be generally useful.

THEORY

Electrodes based on organic ion exchangers

The theory of these ion-exchange electrodes has been developed in detail by Eisenman and his colleagues and is given in Ref. 23: they have derived theoretical equations for the membrane potential and selectivity coefficients. Unfortunately, these equations are only soluble in certain simple cases which do not approximate sufficiently closely to the normally used ion-exchange electrodes for the solutions to be used as any more than approximate guides to electrode performance. It is not yet possible to calculate accurately the selectivity coefficient of an ion-exchange electrode. Thus, the theory will be presented in an abbreviated form. The references will provide a more detailed treatment.

In general terms, the detection limit is set by the equilibrium activity of the determinand in the membrane and the distribution coefficient for the determinand between the membrane and the sample. Similar considerations apply to the derivation of selectivity coefficients. As a simple example, if the response of an ion-exchange membrane, containing the ion-exchanger S^-, to the determinand A^+ suffers interference from B^+ ions, then, as a first approximation, the selectivity coefficient, k_{AB}, is the equilibrium constant for the reaction

$$B^+ \text{ (sample)} + AS \text{ (membrane)} \rightleftharpoons BS \text{ (membrane)} + A^+ \text{ (sample)}$$

and

$$k_{AB} = \frac{K^*_{BS} \, k_{B^+}}{K^*_{AS} \, k_{A^+}}$$

where K^*_{AS} and K^*_{BS} are the stability constants in the membrane for AS and BS and k_{A^+} and k_{B^+} are the partition coefficients of A^+ and B^+ between the sample solutions and the membrane. Such a treatment, which ignores the differing mobilities of A^+ and B^+, and AS and BS in the membrane, frequently allows the selectivity coefficients for the different interferents to be ranked in the correct order.

Hydrogen ion interference, which limits the working range of sample pH, has been discussed in detail for the case of the calcium electrode by Růžička *et al.*[6] The graph of the response of the Orion calcium electrode to a constant activity of calcium ions at different pH values is known to show a dip centred at about pH 4; at lower pH the curve rises steeply as the electrode begins to respond directly to hydrogen ions (Fig. 6.2). There is evidence that the dip is due to a change in the

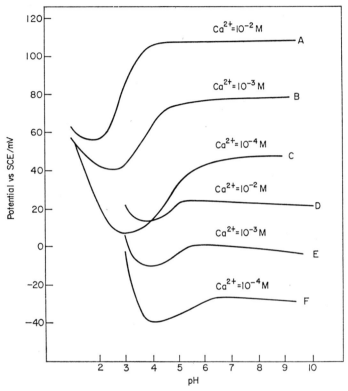

Fig. 6.2   A graph showing the variation with pH of the response of calcium-selective membranes based on a weakly acidic ion exchanger (DOPP) (curves A, B and C) and the Orion active material (DDP) (curves D, E and F). Adapted from Ref. 6

value of the extraction coefficient of the calcium ions (i.e. [CaS_2] (membrane)/ [Ca^{2+}] (solution)) as the predominant species in the membrane changes from $Ca(HS_2)_2$, via species such as $Ca(HS_2)_2HS$ and $Ca(HS)_2HS_2$ to HS. If the less acid complexes such as CaS_2 give rise to lower extraction coefficients than the partially acidic complexes, the potential–pH curve would show the experimentally observed dip before the hydrogen ion response takes over and all calcium ions are displaced from the membrane. Růžička *et al.* showed that the depth of the dip varied with the speed and direction of the pH change; also, by changing to a more weakly acidic ion exchanger they shifted the position of the dip 1–2 pH units towards the acid region (Fig. 6.2).

The role of the solvent in these ion-exchange membranes is complicated; primary requirements are that it should be immiscible with water, exclude ions of opposite charge to the determinand and be nonvolatile. In the case of solid membranes the solution of ion exchanger in the solvent must be capable of forming a homogeneous solution in the inert matrix to give a clear membrane, not exuding solvent.[5] Also, the membrane should remain plastic at room temperature to maintain the mobility of the ion-exchange molecules.

The choice of solvent also has a drastic effect on the selectivity of the membrane. This is clearly shown in the case of the calcium electrode which contains an ion exchanger dissolved in the polar solvent di(n-octylphenyl) phosphonate; if this solvent is changed to the less polar decanol, then the electrode loses its selectivity for calcium ions and senses all divalent ions almost equally. This behaviour is utilized in the 'divalent cation' electrode.

Electrodes based on neutral carrier electrodes

The theory developed for electrodes employing neutral carriers as active material has been much more successful, in predicting precisely the performance of the electrodes, than has the theory for ion-exchange electrodes. Much of the theory was again developed by Eisenman and his colleagues;[23] this and more recent work has been reviewed by Simon and his colleagues.[24,25]

The neutral carriers used in these electrodes are usually macrotetrolide antibiotics, depsipeptide antibiotics or, sometimes, cyclic polyethers (crown compounds). These molecules all have closely knit structures with a central cavity, and tend to favour complexation with ions which are similar in size to the cavity. Thus, by correct choice of carrier and solvent, electrodes have been made with high selectivity for the determinand: also, where no suitable carrier was already available, neutral carriers have been tailored to give the correct cavity size for a particular determinand.[25]

The response limits of these electrodes are set by the same considerations as those for the ion-exchange electrodes. In the prediction of selectivity coefficients the arguments also follow similar lines, but the resultant equations are followed much more closely in practice. For a neutral ligand, S, the controlling reaction is

$$B^+ \text{ (solution)} + AS^+ \text{ (membrane)} \xrightleftharpoons{k_{AB}} BS^+ \text{ (membrane)} + A^+ \text{ (solution)}$$

The expression may now usefully be taken a stage further than before:

$$k_{AB} = \frac{K^*_{BS^+} \, k_{B^+}}{K^*_{AS^+} \, k_{A^+}} = \frac{K_{BS^+} \, k_{BS^+}}{K_{AS^+} \, k_{AS^+}}$$

where K_{AS^+} and K_{BS^+} are the stability constants of AS^+ and BS^+ in the solution and k_{AS^+} and k_{BS^+} are the partition coefficients of AS^+ and BS^+ between the sample solutions and the membrane. For neutral carriers such as the antibiotics, which completely envelop monovalent cations, a good assumption is that

$$k_{BS^+} = k_{AS^+}$$

Hence,

$$k_{AB} = \frac{K_{BS^+}}{K_{AS^+}}$$

It has been shown[24,25] that a close correlation does exist in practice between k_{AB} and K_{BS^+}/K_{AS^+} for both the macrotetrolide antibiotics monactin and nonactin and the depsipeptide antibiotic valinomycin. Such close correlation is shown in some but not all cases by the crown compounds.[27] Divergences are attributed to the inability of the ligand to displace completely the solvent shell of the cation; this may be because the radius of the ligand cavity in which the cation sits is large with respect to the radius of the cation or because the dielectric constant of the solvent in the membrane is too low.

The dielectric constant of the solvent is an important factor influencing the selectivity coefficient. In particular, the preference of a membrane for monovalent ions over divalent ions is enhanced as the dielectric constant drops and vice versa. Thus, in the design of a calcium electrode a solvent of high dielectric constant was chosen to maximize k_{CaNa},[25] whereas for the potassium electrode (Table 6.3) diphenyl ether, a solvent of low dielectric constant, was chosen ($\varepsilon = 3.7$). The dielectric constant of the solvent also affects the electrode response time, as shown by Morf, Lindner and Simon.[28] They studied valinomycin-based potassium electrodes and found that a change of dielectric constant from about 5 to 24 increased the response time at least tenfold.

The performance of the electrodes is improved by the addition of tetraphenylborate ions to the membrane[29,30] to prevent interference from lipid soluble anions such as chloride, thiocyanate and perchlorate. The effect of the tetraphenylborate ions was derived theoretically and demonstrated practically for calcium and potassium electrodes.

PERFORMANCE

Response range and slope

The response ranges of the most important ion-exchange electrodes are given in Table 6.4: most of the data are taken from independent evaluations. Also included in the table are the optimum pH ranges for samples containing a determinand activity well above the limit of detection. The electrodes show concentration hysteresis if used in very strong solutions of determinand for long periods, as the loading of determinand in the membrane is increased. Drift and deviation from the Nernst equation also occur when the concentration of determinand in the sample becomes much larger than in the filling solution.[31]

TABLE 6.4
Response ranges of ion-exchange electrodes at 25 °C

Electrode	Upper useful limit/M	Limit of Nernstian response/M	Limit of detection/M	pH range
Ca^{2+} (a), (b)	1	3×10^{-5}	3×10^{-6}	6–10
(c)	1	3×10^{-6a}	1×10^{-8a}	5–10
NO_3^- (a), (b), (c), (d)	1	1×10^{-4}	5×10^{-6}	3–8
ClO_4^-	10^{-1}	5×10^{-5}	1×10^{-6}	4–10
BF_4^-	10^{-1}	1×10^{-4}	2×10^{-6}	2–12
Divalent cation	1	3×10^{-5}	3×10^{-6}	6–10
CO_3^{2-}	10^{-2}	10^{-7}	10^{-8}	6–9
U (VI) (a)	?	2×10^{-6}	?	4–5
(b)	10^{-1}	1×10^{-4}	1×10^{-5}	2–4
Cl^-	1	3×10^{-4}	1×10^{-5}	3–10

[a] In Ca^{2+} buffer.

The lower pH limits are mostly set by the onset of protonation of the ligands; those ligands containing functional groups which can take part in acid–base reactions are particularly susceptible to this hydrogen ion interference. As previously mentioned, the alkylphosphate ion exchangers used in the calcium-selective electrodes show pronounced pH effects as the predominant species in the membrane are converted into the acid form and the calcium ions displaced.[15] The lower limit of the carbonate-selective electrode is fixed at the pH at which a significant proportion of the carbonate ions are converted to bicarbonate ions, to which the electrode does not respond. The upper pH limits of the calcium and divalent cation electrodes are similarly fixed by the onset of complexation of the determinand, in this case formation of $CaOH^+$ and $MgOH^+$ complexes. For the anion-selective electrodes hydroxyl ion interference is the limiting factor. In the case of the nitrate electrode, Davies et al.[7] observed an intolerable increase in the response time in samples of pH greater than 8.

Similar data for the electrodes based on neutral carriers are presented in Table 6.5. These electrodes show approximately the same response ranges as the

TABLE 6.5
Response ranges of neutral carrier electrodes at 25° C

Electrode	Upper useful limit/M	Limit of Nernstian response/M	Limit of detection/M	pH range
K^+ (a), (c)	1	1×10^{-5}	10^{-6}	3–10
(b)	1	3×10^{-5}	10^{-6}	3–10
NH_4^+	10^{-1}	1×10^{-5}	10^{-6}	5–8
Ca^{2+}	1	3×10^{-5}	10^{-6}	4–10
		1×10^{-7a}		
Ba^{2+}	10^{-1}	1×10^{-5}	10^{-6}	5–9

[a] In Ca^{2+} buffer.

ion-exchange electrodes but, in general, tolerate a wider sample pH range. The upper pH limits of the ammonium and calcium electrodes are set by complexation of the determinand with hydroxyl ions; although the electrodes will respond at higher pH values, the responses will only be to those fractions of the determinands present as NH_4^+ and Ca^{2+}. The lower limits are, as with nearly all the electrodes, dependent on determinand concentration: thus, the lower pH limit of the ammonium electrode in a 10^{-2} M ammonium solution is pH 3, but in a 10^{-4} M solution the limit is pH 6.

Selectivity

In general, the ion-exchange electrodes are less selective than the other types of ion-selective electrodes; hence, an appreciation of the selectivity of an electrode is essential in the development of an analytical method. Unfortunately, the electrodes are not sufficiently well behaved to permit accurate and reproducible measurement of selectivity coefficients and, as previously discussed, attempts to calculate the coefficients theoretically have been largely unsuccessful. The coefficients vary considerably with the activities of primary and interfering ions and show hysteresis. Hence, use of published coefficients to correct the measured response of an electrode for the presence of interferents does not usually produce a satisfactory answer: the published coefficients should only be used as rough guides to the order of magnitude of the interference produced by a particular ion.

For the measurement of selectivity coefficients of ion-exchange electrodes it is more important to use a mixed solution method than for any other class of electrode. Hence, it is regrettable that so many coefficients are still measured by the obsolete separate solution method. The variety of measurement methods used and the irreproducible selectivity behaviour of the electrodes account in part for the range of measured values of selectivity coefficients. A resumé of published data on the calcium and nitrate electrodes is given in Table 6.6. Details of the other, less-used, ion-exchange electrodes are given in abbreviated form in Table 6.7.

From Tables 3.2 and 6.6 it may be seen that there is practically no difference between the selectivity of the liquid and solid forms of the ion-exchange membranes containing calcium di(n-decyl) phosphate, but better than both is the membrane containing calcium di(n-octylphenyl) phosphate. In all cases the most serious interference comes from zinc, ferrous, lead and cupric ions. The interference caused by zinc ions is not sufficiently represented by a simple selectivity coefficient as exposure of a calcium electrode to zinc ions produces a potential shift from which the electrode recovers only slowly;[4] thus, it is safer to assume that the largest of the quoted values of k_{CaZn} is correct.

The three measured values of k_{CaNa} differ widely. This is almost certainly because the selectivity coefficient depends on the activities of calcium and

TABLE 6.6
Selectivity coefficients of ion-exchange electrodes for calcium and nitrate

Electrode	Type	Method of measurement	Interferent	Concn of interferent/M	Selectivity coefficient	Ref.
Ca^{2+}	(a)		see Table 3.2			4
	(a)	mixed soln?	Zn^{2+}		3.2(?)	32
			Fe^{2+}		0.80	
			Pb^{2+}		0.63	
			Cu^{2+}		0.27	
			Ni^{2+}		8×10^{-2}	
			Sr^{2+}		2×10^{-2}	
			Na^+		1.6×10^{-3}	
	(b)	2	Mg^{2+}	4.3×10^{-5} 5×10^{-3}	0.14–0.025	4
			Ba^{2+}	5.0×10^{-3} 5×10^{-2}	7.2×10^{-3} 4.0×10^{-3}	
			Zn^{2+}	5.0×10^{-4} 5×10^{-3}	0.06–0.045	
			K^+	1.0	3×10^{-5}	
			Na^+	1.0	5.8×10^{-5}	
	(c)	sep. soln	Mg^{2+}		2.5×10^{-4}	6
			Sr^{2+}		1.7×10^{-2}	
			Ba^{2+}		2.5×10^{-4}	
			Cu^{2+}		1.6×10^{-2}	
			$Zn^{?+}$		6.0×10^{-2}	
			Cd^{2+}		3.0×10^{-4}	
			Li^+		5.8×10^{-5}	
			K^+		2.0×10^{-6}	
		mixed soln	Na^+		6.3×10^{-6}	
NO_3^-	(a)	mixed soln?	Cl^-		4×10^{-3}	manufacturers' data
			Br		1.1×10^{-2}	
			I^-		25	
			ClO_4		$>10^3$	
	(b)	1	F^-	5×10^{-2}	8.7×10^{-4}	7
			Cl^-	5×10^{-1}	5×10^{-3}	
			I^-	5×10^{-5}	17	
			NO_2^-	5×10^{-2}	6.6×10^{-2}	
			SO_4^{2-}	5×10^{-1}	$<10^{-5}$	
			ClO_3^-	5×10^{-4}	1.66	
			ClO_4^-	5×10^{-5}	800	
	(c), (e)	mixed soln? several	F^-		9×10^{-4}	32
			Cl^-		6×10^{-3} 5×10^{-2}	32–34
		3	Br^-	2.4×10^{-2} 1.4×10^{-3}	9×10^{-2} 1.3×10^{-1}	35

TABLE 6.6—*contd*

Electrode	Type	Method of measurement	Interferent	Concn of interferent/M	Selectivity coefficient	Ref.
		3	I^-	1.2×10^{-2} 3×10^{-6}	15.7–3.2	35
		several	NO_2^-		6×10^{-2} 9×10^{-2}	32, 34
			HCO_3^-		2×10^{-2}	32
			ClO_4^-		10^3	32
			CN^-		2×10^{-2}	32
		3	ClO_3^-	9×10^{-3}– 3.6×10^{-5}	1.14–1.21	2
	(d)	2	F^-	5×10^{-2}	7×10^{-4}	7
			Cl^-	5×10^{-1}	4×10^{-3}	
			I^-	5×10^{-5}	16	
			NO_2^-	5×10^{-2}	6×10^{-2}	
			SO_4^{2-}	5×10^{-1}	3×10^{-4}	
			ClO_3^-	5×10^{-4}	1.66.	
			ClO_4^-	5×10^{-5}	550	

sodium ions in the solutions used to determine it (from Eq. 3.11, $k_{CaNa} = {}_T a_{Ca^{2+}}/a_{Na^+}{}^2$); these activities were presumably different in the three experiments, although, unfortunately, insufficient details are given in the relevant papers to prove this. However, it is clear that when the sodium activity in a sample is high relative to the calcium ion activity, significant interference with the response of the calcium electrode is observed.

TABLE 6.7
Selectivity coefficients of other ion-exchange electrodes

Electrode	Interferent—selectivity coefficient	Ref.
ClO_4^-	I^-—1×10^{-2} to 7×10^{-2}, F^-—3×10^{-4}	32, 35, 36
	Cl^-—2×10^{-4}, Br^-—6×10^{-4} to 1×10^{-3}	
	NO_3^-—2×10^{-3} to 4×10^{-3}, HCO_3^- 4×10^{-4}, OH^-—1	
	MnO_4^-, IO_3^-, $Cr_2O_7^{2-}$, ReO_4^{2-} and CNS^- also interfere	37, 38
BF_4^-	OH^-—10^{-3}, I^-—20, NO_3^-—0.1, F^-—10^{-3}	32
	Cl^-—10^{-3}, Br^-—4×10^{-2}, HCO_3^-—4×10^{-3}	
	Acetate—4×10^{-3}	
Divalent cation	Ca^{2+}—1, Mg^{2+}—1, Zn^{2+}—3.5, Fe^{2+}—3.5, Mn^{2+}—2	32, 39
(Ca^{2+} and Mg^{2+})	Cu^{2+}—3.1, Ni^{2+}—1.4, Ba^{2+}—0.94, Sr^{2+}—0.54	
CO_3^{2-}	SO_4^{2-}—1.5×10^{-4}, HPO_4^{2-}—2.6×10^{-4}, Cl^-—1.9×10^{-4}	11
	NO_3^-—0.29, ClO_4^-—25, CH_3COO^-—2.6×10^{-2}	
	Borate—4.8×10^{-2}, Hydrogen phthalate—83	
Cl^-	ClO_4^-—32, I^-—6 to 24, NO_3^-—2.6 to 4.4	32, 35
	Br^-—1.6 to 3.4, OH^-—1.0, Acetate—0.32	
	HCO_3—0.19, F^-—0.10	

For the nitrate electrodes there is again little to choose between the different types. The ions which must be particularly excluded from samples are perchlorate, chlorate and iodide: fortunately, these ions are not frequently present in samples for nitrate analysis. However, the selectivity coefficients for chloride, bicarbonate and nitrite ions are often inconveniently large and samples must be pretreated to remove these interferents, as described in the section on applications.

The chloride electrode based on the ion exchanger is much less selective to chloride with respect to ions such as nitrate, bicarbonate and acetate than the electrode based on silver chloride (see Table 5.7). However, it does have the advantage that it can tolerate the presence of traces of sulfide and iodide in samples in which the other electrode cannot be used, although frequently samples may be treated to remove iodide and sulfide before measurement. It is nearly always preferable to use one of the electrodes based on inorganic salts for chloride measurement if possible.

The divalent cation electrode shows approximately equal sensitivity to all divalent cations and is thus very suitable for the measurement of water hardness $(Ca^{2+} + Mg^{2+})$.

The selectivity coefficients of the electrodes based on neutral carriers are much more amenable to precise determination than those for the ion-exchange electrodes; in particular, they vary less with the relative concentrations of determinand and interferent. A comparison of the selectivity coefficients of the various types of potassium electrode based on valinomycin is given in Table 6.8:

TABLE 6.8
Selectivity coefficients of valinomycin-based potassium electrodes

Interferent	Concn of interferent/M	Membrane (a) Selectivity coefficient	Ref.	Membrane (b) Selectivity coefficient	Ref.	Membrane (c) Selectivity coefficient	Ref.
Na$^+$	0.1	9.6×10^{-5}	15	6.0×10^{-5}	15		
		2.5×10^{-4}	14			3.3×10^{-4}	16
		7×10^{-5}	30				
NH$_4^+$	0.1	1.0–1.6×10^{-2}	14, 15	1.3×10^{-2}		2.3×10^{-2}	
Cs$^+$	0.1	4.7×10^{-1}	15	4.7×10^{-1}			
		4.0×10^{-1}	14			3.4×10^{-3}	
Rb$^+$	0.1	4.7	15	4.7			
		1.9	14			1.9	
Li$^+$	0.1	2×10^{-4}	14, 15	2×10^{-4}		6.3×10^{-4}	
Ca^{2+}	0.1	4×10^{-5}	15	5×10^{-4}			
		2×10^{-4}	14, 40			9×10^{-4}	
Mg^{2+}	0.1	4×10^{-5}	15	4×10^{-5}			
		2×10^{-4}	14, 40			6×10^{-4}	
Ba^{2+}	0.1	1×10^{-4}	14	1×10^{-4}			
		6×10^{-5}	14			7×10^{-4}	

data on the interference caused by one or two other ions are included in the references quoted in the table. The selectivity coefficients for other electrodes based on neutral ligands are given in Table 6.9.

TABLE 6.9
Selectivity coefficients for other neutral carrier electrodes

Electrode	Interferent—selectivity coefficient	Ref.
NH_4^+	K^+—1.2×10^{-1}, Na^+—2.0×10^{-3}, Rb^+—4.3×10^{-2} Cs^+—4.8×10^{-3}, Li^+—4.2×10^{-3}, Mg^{2+}—2×10^{-4}, Sr^{2+}—1×10^{-1}, Ba^{2+}—9×10^{-1}, Zn^{2+}—6×10^{-4}	17
Ca^{2+} ([interferent]= 10^{-2} M)	H^+—4.0×10^{-5}, Na^+—3.6×10^{-4}, NH_4^+—7.9×10^{-6} Zn^{2+}—1.4×10^{-4}, Mg^{2+}—3.8×10^{-5}, Sr^{2+}—7.1×10^{-3}	18
Ba^{2+}	Sr^{2+}—2×10^{-3}, Co^{2+}, Mg^{2+}, Ni^{2+}, Ca^{2+}, Zn^{2+}, $Fe^{2+} < 1 \times 10^{-4}$, K^+—8×10^{-3}, Na^+—2×10^{-4}, NH_4^+—6×10^{-4}, Li^+—2×10^{-4} Cu^{2+} poisons the electrode	19

Table 6.8 shows that the differences in selectivity are negligible between the liquid membrane with valinomycin in diphenyl ether and the solid membrane with the valinomycin plus solvent in PVC. This is confirmed by the work of Mascini and Pallozzi.[2] The electrodes made from valinomycin in silicone rubber (without solvent) show marginally worse selectivity except with respect to interference by caesium ions. The major interferents are the relatively unimportant rubidium and caesium ions. Hammond and Lambert[41] have reported interference to the liquid membrane potassium electrode from surface-active compounds, such as hexadecyltrimethylammonium bromide and alkyldiethanolamines containing more than five carbon atoms in the alkyl chain; the authors concluded that interferents must possess both a positive charge and surface active properties. This interference was attributed to dissolution of the interferent in the membrane solvent; adjustment of the pH of samples suppressed the positively charged form of the interfering molecule and also suppressed the interference.

The ammonium electrode has better selectivity for ammonium ions over other monovalent cations than the ammonium-selective glass electrode. However, its performance does not approach the high selectivity of the ammonia-sensing membrane probe and thus it has not been widely used.

The new calcium electrode based on the neutral carrier[18] shows much higher selectivity than the calcium ion-exchange electrode both with respect to other divalent ions such as magnesium and zinc and also with respect to monovalent cations. This high selectivity together with the high sensitivity are important advances in the development of calcium electrodes.

Stability and reproducibility

Růžička et al.[6] report that the standard potential of an Orion calcium electrode was 83.7 ± 2.49 mV over the first 3 weeks of its use; after three weeks the performance deteriorated and the standard potential drifted, so that recharging of the membrane was necessary. The stability of their electrode (Table 6.2, type c) was similar but was maintained over a longer period; thus, over 7 weeks the standard potential was 60.56 ± 2.75 mV. Moody et al.[4] found that an electrode, made from the Orion ion exchanger incorporated in PVC, drifted by as much as 6 mV per day in 10^{-2} M calcium chloride solution; however, the cumulative drift over 18 months was only $+20$ mV.

In their evaluation of the Orion nitrate electrode, Potterton and Shults[34] found that electrode potentials were reproducible to ± 0.8 mV over 3 h and ± 2.9 mV over 5 days. Exposure to interferents such as iodide left the electrode unstable for several minutes. With an electrode made from the Orion exchanger in PVC, Davies et al.[7] noted that electrode potentials in a fixed concentration of nitrate varied by an inconveniently large amount (± 4 mV) during 24 h and the standard potential drifted by $+5$ mV over 2 weeks.

Fiedler and Růžička[15] found that the standard potential of a conditioned Philips potassium electrode drifted by 22 mV in 1 week, whereas their 'Selectrodes' with PVC membranes showed negligible drift in 3 weeks. However, Pioda et al.[14] found that the Philips potassium electrode drifted over a maximum range of ± 1.4 mV in 6 days; the silicone rubber electrode of Pick et al.[16] drifted less than 0.1 mV in 5 days. The reproducibilities at the 95% confidence level of the Philips electrode and the silicone rubber electrode were both ± 0.5 mV ($\pm 2\%$).[16]

Fleet and Rechnitz[42] demonstrated the sensitivity of liquid ion-exchange electrodes—in particular, the Orion calcium electrode—to the rate of flow of the sample solution across the membrane in a closed flow system. At concentrations above 10^{-4} M Ca^{2+}, the electrode potential increased steadily as the flow velocity was increased until the change was about 2 mV at a velocity of 150 cm s^{-1}: potential measurements could be made with an accuracy of ± 0.2–0.4 mV. Above the velocity of 150 cm s^{-1}, the potential response became erratic, presumably owing to distortion of the membrane by the high pressure of the sample. At calcium concentrations below 10^{-4} M the electrode response became very dependent upon velocity and showed a sharp increase of about 4 mV as the flow started; the precision of potential measurements was poor at these concentrations.

Response time

The response times of these electrodes are much longer than the response times of the glass electrodes and the electrodes based on inorganic salts. In general, the ion-exchange electrodes with solid membranes respond slightly faster than those with liquid membranes, whereas, for the electrodes based on neutral carriers, the versions with liquid membranes are much faster.[43] Moody and Thomas[39] have summarized response time data for the electrodes and Morf et al.[28] and Markovic and Osburn[44] have treated the response times theoretically.

The Orion calcium electrode, which has a liquid membrane, has been the subject of a response time study by Fleet, Ryan and Brand.[45] They measured the times taken for 95% of the total potential change to occur after activity changes created by rapidly injecting strong determinand solution into vigorously stirred samples. In pure calcium solutions the response time (t_{95}) for decadic activity changes over the range 10^{-4}–10^{-1} M was 2.3 ± 0.2 s: this value is similar to those obtained by other workers. However, in solutions containing cations in addition to calcium ions some interesting effects were observed which allowed these cations to be separated into three main groups. First, there are those ions, such as sodium, for which the selectivity coefficient is small and which produce a negligibly small effect on the response times. Second, there are ions, such as copper, lead and cadmium, for which the selectivity coefficient is large but which still do not affect the response times. Finally there is a group of ions, including the alkaline earths magnesium, strontium and barium and also zinc, for which the selectivity coefficients are intermediate but produce a significant slowing of the electrode response, even if their activity is insufficient to interfere with the final measured potential. Fleet et al.[44] attribute the slowing of the response caused by this last group of ions to the slower rate of reaction of these ions with the ion exchanger and the slower transport of the ions across the membrane–solution interface in comparison with calcium ions. Thus, the rate of attainment of the following equilibrium is slowed:

$$AS_2 \text{ (membrane)} + B^{2+} \text{ (soln)} \rightleftharpoons BS_2 \text{ (membrane)} + A^{2+} \text{ (soln)}$$

where A^{2+} is the determinand, B^{2+} the interferent and S^- the ion exchanger. In the reported experiments, with relatively high levels of interferents, the presence of magnesium doubled the response time and barium produced a tenfold increase.

Potterton and Shults[34] studied the response time of the Orion nitrate electrode and showed it to be comparable to the response time of a conventional pH glass electrode in pH buffers, under the conditions of measurement. The nitrate electrode took about 15 s to come to within 1 mV of its final equilibrium potential after an activity change from 10^{-2} to 10^{-3} M nitrate or from 10^{-4} to 10^{-3} M nitrate. The response times reported by Davies et al.[7] for the PVC electrode for the final equilibrium value to be reached (to ± 0.2 mV) after an activity change (t_{100}), ranged from approximately 2.5 to 5 min in the range 10^{-1}–10^{-4} M nitrate, but were extended to 30 min at 10^{-5} M; in these experiments the electrode was transferred between beakers to change the concentration and, hence, response times as short as those which would be observed in a flowing system are not expected.

Morf et al.[28] investigated the response of potassium electrodes with membranes of valinomycin dissolved in dipentylphthalate and incorporated in

PVC. They showed that for changes from 10^{-3} to 10^{-2} M potassium, in experiments in which the potassium ion activity was changed *in situ* in rapidly stirred solutions, the electrode came to within 0.2 mV of its final potential in about 15 s. In the same system but with slow stirring the equivalent response time was more than 6 min. Moody and Thomas[39] report that the presence of caesium, magnesium, strontium or barium ions in large concentrations slow the response.

Temperature effects

Negus and Light[46] determined the nonisothermal temperature coefficient of the Orion divalent cation electrode and found it to be $+0.52$ mV K^{-1}.

Potterton and Shults[34] studied the effect of temperature on the calibration curve of the Orion nitrate electrode. The curves at 35 °C and 45 °C were of similar form to those obtained at 25 °C with slopes 4–5 % less than the theoretical Nernstian value. Because of the increased solubility of the ion exchanger in water at the higher temperatures, the deviations of the curves from linearity were more pronounced than at 25 °C.

In their study of the Orion potassium electrode Eyal and Rechnitz[26] found that if the electrode was cooled to 5 °C, the membrane froze and ceased to respond. However, when the experiment was later repeated by Fiedler and Růžička,[15] with the similar Philips potassium electrode, no such effect was observed and the electrode responded quite satisfactorily with its selectivity maintained. But when the electrode was warmed up again to 25 °C, a slight deterioration in performance was noted. The electrode they[15] developed (Table 6.3, type b) did not deteriorate after cooling and worked satisfactorily at 5 °C.

Lifetime

It is virtually impossible to present a realistic picture of the expected lifetime of an ion-exchange electrode, as it depends so much on the type of usage the electrode gets. Unfortunately, no standard usage test has yet been developed which would enable electrode lifetimes to be tested and compared, and the acceptance of any such test is improbable, as the electrodes are commonly used in so many types of application that no usage can be called standard. Thus, the situation remains that reference to the papers published in favour of the different designs allows no worthwhile comparison to be made between them; statements about electrode lifetime are seldom sufficiently qualified by experimental details to make them at all useful.

Some of the parameters affecting the lifetime of electrode membranes were discussed in the introduction to this chapter but others may also be mentioned. The life of an electrode may be maximized by carefully pretreating samples to ensure that interferents, and substances which are appreciably soluble in the membrane, are removed before the samples are presented to the electrode. Also, electrodes with liquid membranes should be stored in a dilute solution of the determinand. If the electrode is left in air for a long period, volatilization of the solvent in the membrane may shorten the membrane life; the same effect leads more slowly to the failure of some solid membranes.[2] A further safeguard is to avoid using the electrodes at high temperatures and by so doing minimize the solution of the active material of the membrane in the samples.

Among the most careful studies of electrode lifetime is that of Růžička *et al.*[6] on calcium electrodes. Between measurements they stored the electrodes in a solution containing 10^{-2} M calcium ions and 10^{-1} M sodium ions. The Orion calcium electrode lasted 3 weeks before needing renewal, whereas these authors' calcium electrode with a solid membrane lasted more than 7 weeks; it is not clear whether this difference is attributable to the form of the membranes or the different solubilities of the ion exchangers. However, as shown in Table 6.1, lifetimes of approximately 1–2 months are common.

APPLICATIONS

Calcium electrode

The calcium electrode has made most impact in the field of biomedical analysis. In this field, great attention must be paid to experimental detail if satisfactory results are to be obtained in those applications, which are usual in this kind of work, in which the electrode is used to monitor small variations in calcium ion activity over a narrow concentration range. On account of the logarithmic response of ion-selective electrodes, high precision is always difficult to achieve, but it is particularly difficult with electrodes for divalent cations such as calcium.

The ability of the electrode to measure only ionized calcium activity, in the presence of various bound forms of the element, has made possible wide-ranging research into the role of calcium ions, which are the only physiologically active calcium species, in various physiological processes. In addition to such research uses, the electrode is also widely applied to the routine analysis of calcium in serum and plasma. For these analyses the electrode is usually incorporated into a flow-through system specially designed to prevent loss of carbon dioxide (and, hence, pH change with consequent change in the proportion of calcium ions unbound) and to allow analysis of small samples. The constant flow rate of the solution past the electrode helps to stabilize both the calcium electrode readings and the liquid junction potential at the reference electrode, and also to minimize the deposition of protein on the membrane surface. The system developed by Orion has been widely used but has proved to be rather temperamental. A newer system developed by that firm is simpler to use and should be more satisfactory, as it includes, among other improvements, good temperature control: this latter feature is essential both to prevent electrode drift and to ensure constancy of the proportion of unbound calcium ions in the samples.

Frequent calibration of a calcium electrode used in biological fluids is essential to maintain accuracy, since drifts of approximately 2–5 mV per day are observed even when a flow-through system is used:[47] such drifts must be compared with a total potential change of only about 1 mV between the electrode potential in successive samples of serum containing calcium activities differing by the relatively large amount of 10%. A further problem is that transfer of the electrode from aqueous standards to samples containing protein causes a shift in the calibration curve by 1–2 mV.[48] Various methods have been used to minimize this 'protein shift', including running aqueous standards before and after each sample and adding trypsin and triethanolamine to the standards; these methods are discussed in Ref. 48.

Covington and Robinson[49] have proposed calcium ion activity standards for the standardization of electrodes for the analysis of calcium in blood serum. The solutions, which are listed in Table 3.2, have a background which is approximately the same as that of blood serum.

Moore,[47] in a very useful review of his own and other work with the Orion calcium system, gives details of many varied experiments which are interesting because of both the methods used and the results obtained. Further work has been reviewed in Refs. 48 and 50 and in the biennial reviews in *Analytical Chemistry*.

The calcium electrode also finds many uses outside the area of biomedical analysis. A simple application is the analysis of calcium in water supplies: Hulanicki and Trojanowicz[51] have suggested a useful direct method for this determination. They developed a 'constant complexation buffer' for addition at a volume ratio of 1:1 to both samples and standards. This ensures that the correct pH is achieved (pH 9.1), magnesium ions are masked, the ionic strength is fixed and a constant proportion of the calcium ions is complexed. The buffer consists of:

0.02 M ammonium hydroxide
0.02 M ammonium chloride
0.4 M potassium nitrate
0.04 M acetylacetone
0.02 M iminodiacetic acid

With this buffer, determinations of calcium in the range 20–800 mg Ca^{2+} per litre were performed with an average accuracy of 2–4%.

Moody *et al.*[4] measured the calcium content of Cardiff tap-water on ten days both with a calcium electrode and by an EDTA titration procedure. The results from the methods were in close agreement in all cases and the means of the results were 31.62 mg Ca^{2+} per litre for the EDTA titration and 31.51 mg Ca^{2+} per litre for the liquid ion-exchange electrode.

Muldoon and Liska[52] have reported the determination of ionized calcium in milk using both a direct calcium electrode method and an ion-exchange/ titrimetric method. The authors found close agreement between the results of the two methods in the analysis of both pasteurized and sterilized milk, but for the raw milk there was a significant difference: the ion-exchange method gave the concentration of ionized calcium as $2.52 \pm 0.15 \times 10^{-3}$ M, whereas the electrode gave $2.71 \pm 0.01 \times 10^{-3}$ M. The authors attributed this difference to an error in the ion-exchange method resulting from the strong pH dependence of the properties of the ion exchanger coupled with microbial activity in the milk perhaps causing pH changes: these pH changes were not demonstrated.

A standard addition method for the determination of calcium in beer has been described.[53] In this case pH adjustment is necessary to raise the sample pH to the correct working range and, because an excess of calcium complexing agents such as proteins are present, the standard addition procedure is essential.

Several workers have reported the use of the calcium electrode in the potentiometric titration of calcium ions with EDTA. In particular, Hadjiioannou and Papastathopoulos[54] determined both calcium and magnesium, and calcium ions alone, using both semi-automatic and automatic procedures. They found it necessary to ensure that in the titration of calcium and magnesium ions the pH of the sample should be greater than 8 at the end-point; a glycine buffer giving an initial pH of 9.7 gave the largest end-point break. For the titration of calcium alone the pH of the solution at the end-point was in the range 11.2–12.0. The authors were able to determine calcium ions in the concentration range 40–320 mg Ca^{2+} per litre (1–8 $\times 10^{-3}$ M) and magnesium ions in the range 30–200 mg Mg^{2+} per litre (1.2–8.3 $\times 10^{-3}$ M) with average errors of 0.2% and 0.3%, respectively (semi-automatic procedure) and 0.4% and 0.7% (automatic procedure).

Fleet and Rechnitz[42] used a calcium electrode with a liquid membrane, in a continuous flow system, to measure the rates of the reaction of calcium ions with lactate, malate, tartrate and gluconate ions and with the magnesium–EDTA complex. With the system developed by the authors, the calcium activity could be measured at fixed times after the reactants were mixed. The shortest reaction time before measurement was 10 ms; hence, relatively fast reactions could be studied.

The calcium content of soda-lime used for making glass has been measured with a calcium electrode by Knupp.[54] He adjusted the pH of the acid-digested samples to 9 by addition of ammonium hydroxide and then determined the calcium by direct potentiometry and compared the results with those from a CDTA titration procedure. The potentiometric results tended to be low by up to 10%. This may partly be attributed to slow electrode response and, hence, insufficient time being allowed for the electrode to equilibrate, and also to non-Nernstian behaviour of the electrode, which had a slope of only 21.8 mV decade^{-1}; it is essential to ensure that an electrode used for an analysis is responding correctly if accurate and reproducible results are required.

Samples to be analysed with the calcium electrode usually need to be buffered to raise the pH to between 6 and 10. A buffer should be used which does not complex the calcium ions (thus, phosphate buffers are unsuitable) or interfere with the electrode response. The addition of sodium ions, either for ionic strength adjustment or in buffers, should be avoided whenever possible to avoid interference from sodium ions. Suitable buffers are the ammonia/

ammonium buffer, triethanolamine/hydrochloric acid buffer and the 'tris'-hydrochloric acid buffer. For most applications single junction calomel or silver/silver chloride reference electrodes are suitable.

Divalent cation electrode

Very few important papers have appeared on the application of this electrode despite the fact that, in the U.K. at least, the sales of this electrode are at least as great as those of the calcium electrode. In general, the same considerations apply to the use of this electrode as have already been discussed for the calcium electrode.

Cheng and Cheng[56] have described a method for determining magnesium in raw water and tap-water using this electrode. Samples are treated with a buffer containing diethanolamine, to fix the pH at 7.0 ± 0.2, and 2,2'-ethylene-dioxybis[ethyliminodi(acetic acid)] (EGTA) as masking agent for other ions (including calcium ions). The method gave good results in the range 10^{-1}–10^{-5} M magnesium ions, which agreed with results obtained by atomic absorption spectroscopy.

The divalent cation, or water hardness electrode, is almost exclusively and quite widely used for the measurement of water hardness (the sum of the activities of divalent cations, primarily calcium and magnesium, in the water). For the continuous monitoring of the hardness of natural and potable water, the electrode has been used in industrial analysers. In the E.I.L. analyser the sample is buffered by the addition of one volume of reagent to every nine volumes of sample before being pumped past the electrode. It has been shown[1] that a suitable buffer is a 0.25 M 'tris' solution adjusted to pH 8.3 by addition of 5 M hydrochloric acid: for use with very hard waters adjustment to pH 9.5 is more satisfactory. An alternative, cheaper, buffer for use in the laboratory is 0.1 M triethanolamine–hydrochloric acid at pH 8; this buffer is not as satisfactory in continuous analysers because it does not prevent so effectively the build-up of precipitates in the tubes.

Nitrate electrode

The nitrate electrode is one of the most popular ion-selective electrodes, partly because all other methods for nitrate analysis are unsatisfactory for one reason or another. But despite the wide usage, primarily in the analysis of nitrate in waters, there is no agreement in the literature on an optimum method. Some authors pretreat samples, some do not; some recommend one type of reference electrode, others another type; and insufficient evidence is presented to suggest that any one method is better than the others.

The application of the electrode in untreated samples is limited primarily by chloride and bicarbonate interference, although variations in the activity coefficient of the nitrate ions can also impair the accuracy. The effect of these interferences may largely be allowed for if the background composition of the samples is known,[33] but usually it is more satisfactory and simpler to pretreat the samples to remove the interferents and to stabilize the activity coefficient. In general, chloride ions interfere, if present in greater than a tenfold excess over nitrate ions, but may be removed by treatment with silver ions, by addition of silver fluoride or silver sulfate solution. The bicarbonate ion interference may be simply removed by adjusting the pH to approximately 4, thus converting the bicarbonate ions to carbon dioxide. In abnormal samples containing high concentrations of nitrite ions, sulfamic acid should be added to prevent interference. For use in continuous analysers for monitoring nitrate in natural and potable waters, a buffer which has been found[1] to give good results when added to samples at a volume ratio of 1:10 is 0.1 M potassium dihydrogen phosphate + 10 g disodium EDTA per litre; this buffer gives a pH of about 4.5.

The choice of bridge solution in the reference electrode, to be used with the nitrate electrode, has been considered by several authors and the following conclusions may be drawn. The use of a bridge solution containing a large concentration of nitrate (e.g. Orion solution No. 90-00-01) or chloride (e.g. saturated potassium chloride solution) should be avoided. Furthermore, if the samples are pretreated with excess silver ions, even relatively low concentrations of chloride in the bridge solution will cause eventual blockage of the junction. If samples are not pretreated with silver ions, the most suitable reference electrode is probably either a double junction reference electrode with a 0.1 M potassium chloride bridge solution (as mentioned in Chapter 2) or a mercury/mercurous sulfate reference electrode with a 1 M or saturated sodium sulfate bridge solution. If the samples are pretreated with silver ions, the most suitable bridge solution is a portion of the pretreatment solution diluted to the same concentration as in the samples. In a method recommended by Orion[57] the pretreatment solution to fix the ionic strength and remove chloride ions consists of 0.8 M $KF \cdot 2H_2O$ and 0.2 M AgF; one volume of this solution is added to every 100 volumes of both samples and standards. The solution, diluted one-hundredfold, is also used as bridge solution in the double junction reference electrode. A drawback of this method is that silver fluoride is very expensive.

When a chloride-containing bridge solution is used with samples not treated with silver ions, it is advisable to make sure that the reference electrode is well away from the nitrate electrode in the samples and to keep the reference electrode in the samples for as short a time as possible. In the more dilute nitrate samples a liquid junction with a low flow rate should be used. All these devices to avoid contamination of the sample by the bridge solution are unnecessary when a flow system with the reference electrode downstream from the nitrate electrode is used.

Milham et al.[58] determined nitrate in powdered plant material, soils and various waters, using the nitrate electrode, and compared the results with those from a modified Devarda method; the correlation between the methods was very close (correlation coefficient, 0.99). Slightly higher results for nitrate in the plant extracts were obtained from the Devarda method, but the differences were consistent with the expected increase of nitrate caused in this method by the hydrolysis of organonitrogen compounds. For river water samples the mean of ten replicate determinations was 4.11 mg NO_3^- nitrogen per litre with a standard deviation of 0.064 mg l^{-1}: the figure given by the Devarda method was 4.12 mg l^{-1}. At a much higher concentration the mean of ten replicate determinations of nitrate in cotton petiole was 17 100 mg NO_3^- nitrogen per kilogram with a standard deviation of 260 mg kg^{-1}; an identical figure was given by the Devarda method. For the analysis of soils, the soil extract was decanted and centrifuged before measurement. In all cases an acidic buffer consisting of 0.01 M aluminium sulfate, 0.01 M silver sulfate, 0.02 M boric acid and 0.02 M sulfuric acid adjusted to pH 3 with 0.1 M sulfuric acid was added to the samples. Milham[59] showed later that by using a flow-through system, with better control of temperature and stirring rate (200 rev min^{-1}), the average coefficient of variation of the results could be reduced from the previous value of 1.5% to 0.6% for samples and 0.4% for standards. In this system the electrode required 30 s to reach equilibrium.

The nitrate electrode has also been used for the analysis of waters and effluents by Bunton and Crosby,[60] who obtained recoveries ranging from 96% to 106% for samples in the range 2.5–25 mg NO_3^- nitrogen per litre; although these authors buffered the samples with potassium dihydrogen phosphate solution, they made no attempt to remove chloride. They found, however, that the electrode method agreed well with a range of other methods in direct determinations and gave consistently reliable recoveries even with sewage effluents, for which the other methods all gave low recoveries.

Forney and McCoy[61] developed a flow-through electrode unit, shown in Fig. 6.3, for use with the nitrate electrode in a system for measuring particulate atmospheric nitrate. The unit provides temperature control to $\pm 0.5\ °C$, good stirring without bubble formation, even for 4 h at 38 °C, and precise liquid level control; the measurement chamber is in the shape of an inverted cone. A fluoride electrode is used as reference electrode and the samples are doped with fluoride ions, as proposed by Manahan[62] and discussed in Chapter 3; the objections[63] to this procedure do not here apply. The system uses a 10^{-2} M silver fluoride collection solution to remove halide and sulfide interferences. The monitor yielded results comparable to those from more laborious standard techniques.

In samples where the interferents are difficult to remove, nitrate may be measured, after reduction to ammonia, with an ammonia probe as discussed in Chapter 7.

Potassium electrode

The potassium electrode based on valinomycin has been used in several studies of biological systems. Special microelectrodes have been prepared, from droplets of ion-exchange solution trapped in the ends of capillaries, and used for the measurement of intracellular potassium activity in living cells. The preparation and application of these microelectrodes has been discussed by Walker.[64]

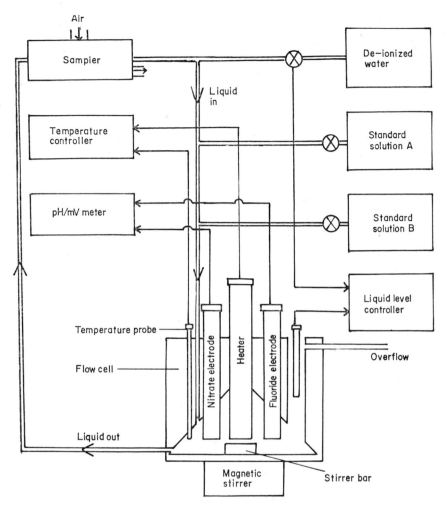

Fig. 6.3   A schematic diagram of a flow-through cell and associated instrumentation to monitor particulate atmospheric nitrate[60]

Because of their very high impedance it is essential that the input impedance of the pH meter to which the electrodes are connected be greater than $10^{13}\ \Omega$. The diameters of the electrode tips, which are made of glass, may be as small as 0.5 μm; they are coated with an organic silicone compound so that the droplet of ion exchanger is retained. Such microelectrodes have been used to study the potassium activity in *Aplysia* neurons.[64]

An example of the use of more conventional potassium electrodes in biological material is the measurement of potassium uptake by brain mitochondria from rats.[65] Protein interference was insufficient to prevent the successful use of a silicone rubber-based electrode. As with the applications of the calcium electrode in biological media, the relatively small variations in determinand activity which have to be measured require careful attention to experimental detail if accurate and precise results are to be obtained.

The reference standards[49] for the determination of potassium in blood serum were previously mentioned in Chapter 3.

Sekerka and Lechner[66] have described the simultaneous automatic determination of potassium, sodium and ammonium ions in natural and waste waters. The potassium method, based on a potassium electrode, was compared with a flame photometric method. Generally, the standard deviations of results were approximately 5% of the true potassium activity down to 1 mg K^+ per litre, but 55% at 0.1 mg K^+ per litre. The correlation of the two methods was very good: the relative errors of the potentiometric method lay in the range $\pm 2\%$. Samples containing more than 10 mg K^+ per litre were analysed at the rate of 20 per hour and more dilute samples at 10 per hour. All samples were treated with an ionic strength adjustment buffer consisting of an alkaline solution of sodium chloride.

Other methods employing the potassium electrode based on valinomycin may be simply developed, as the electrode is relatively well-behaved, and has a wide working pH range and good selectivity. In general, it is to be preferred to the potassium-selective glass electrode because of its greater selectivity. The mercury/mercurous sulfate reference electrode is suitable for use with this electrode; alternatively, a double junction reference electrode with a lithium trichloro-acetate filling solution has been suggested.[57]

Perchlorate and fluoroborate electrodes

Very few applications of these two electrodes have been published and no new principles are involved in their use.

The tetrafluoroborate electrode was developed by Carlson and Paul[10] for the determination of boron; the method was shown to be satisfactory for the analysis of tap-water but a wide variety of other material may also be analysed. The boron in the solution to be analysed was isolated and concentrated on a boron-specific resin and converted to tetrafluoroborate with hydrofluoric acid; the tetrafluoroborate ions were eluted with sodium hydroxide and the solution acidified for measurement by passage through a cation exchange column in the hydrogen form. The tetrafluoroborate ion activity in the final solution was then measured directly. The method gave results for tap-water samples in excellent agreement with those given by the curcurmin method. The calibration range for the method presented was 10^{-5}–2×10^{-3} M boron, but by appropriate adjustment the range could be altered within the limits set by the response range of the electrode. The reference electrode in the cell should have a plastics body as the samples attack glass.

The perchlorate electrode is used almost exclusively for potentiometric titrations. Baczuk and DuBois[37] have described a method involving the titration of perchlorate samples, such as solid propellants, with tetraphenylarsonium chloride solution adjusted to a pH of 4–7: replicate analyses of three types of sample gave excellent agreement with the titanium chloride method and an overall standard deviation on the results of 0.073%. To prevent potassium perchlorate blocking the liquid junction of the reference electrode, as would occur with a potassium chloride bridge solution, a double junction reference electrode with an ammonium nitrate bridge solution was used.

Chloride electrode

The chloride electrode based on an organic ion exchanger is applied almost exclusively in the analysis of samples containing irremovable traces of sulfide or iodide, both of which interfere strongly with the chloride electrodes based on inorganic salts. Such applications are rare.

Walker,[64] prepared, in addition to the calcium microelectrodes already mentioned, chloride microelectrodes from chloride ion-exchange solutions and studied intracellular chloride activities with them.

Carbonate electrode

The carbonate electrode developed by Herman and Rechnitz[11] has been used by these authors in a continuous flow system for the direct determination of carbon dioxide in human serum samples[67] at the rate of 20 per hour. The samples were buffered with a 'tris'-hydrochloric acid buffer to give a pH of about 8.4: variations in sample pH and chloride activity, within the normal ranges in serum, were shown to produce negligible errors. The standard deviation of single measurements was found to be 1.2 meq CO_2 per litre and the results showed close correlation with a standard continuous flow colorimetric method. The method conveniently avoids the troublesome separation, by volatilization, of the carbon dioxide from the samples before measurement.

Ammonium electrode

Applications of the ammonium electrode based on nonactin/monactin have been the subject of few published papers. In general applications it competes with the much cheaper, less troublesome but less selective ammonium glass electrode and the much more selective ammonia probe. It is most suited to the analysis of samples, such as biological materials, which cannot withstand the pH change required for analysis with the ammonia probe, and contain too much sodium or potassium for the glass electrode to be useful. Microelectrodes could be constructed from the ion exchanger for studies in living material.

The ammonium electrode has been compared with the ammonia probe both in general terms[68] and in the analysis of boiler feed-water.[69] The properties of the electrode were fully investigated[68] and the high selectivity coefficients for the alkylammonium salts were measured; the electrode was shown to be extremely selective to the tetrabutylammonium ion. The authors apparently preferred the ammonia probe for the determination of ammonia and nitrate in water samples. In the boiler feed-water analyses[69] either sensor could be used but each suffered one disadvantage: the ammonium electrode had a short lifetime and the probe suffered interference from volatile amines in the samples. The reproducibilities of the electrode and probe methods were both about 1–2% (mean deviation from the mean) with errors in readings of about 2%.

REFERENCES

1. C. L. Jamson, unpublished work in the E.I.L. laboratory.
2. M. Mascini and F. Pallozzi, *Anal. Chim. Acta* **73**, 375 (1974).
3. J. W. Ross, *Science, N.Y.* **156**, 1378 (1967).
4. G. J. Moody, R. B. Oke and J. D. R. Thomas, *Analyst* **95**, 910 (1970).
5. G. H. Griffiths, G. J. Moody and J. D. R. Thomas, *Analyst* **97**, 420 (1972).
6. J. Růžička, E. H. Hansen and J. C. Tjell, *Anal. Chim. Acta* **67**, 155 (1973).
7. J. W. Davies, G. J. Moody and J. D. R. Thomas, *Analyst* **97**, 87 (1972).
8. J. W. Ross, U.S. Patent No. 3,483,112 (9th December 1969).
9. A. Hulanicki, R. Lewandowski and M. Maj, *Anal. Chim. Acta* **69**, 409 (1974).
10. R. M. Carlson and J. L. Paul, *Anal. Chem.* **40**, 1292 (1968).
11. H. B. Herman and G. A. Rechnitz, *Anal. Chim. Acta* **76**, 155 (1975).
12. J. R. Entwistle and T. J. Hayes, *Proceedings* of the I.U.P.A.C. International Symposium on Selective Ion-Sensitive Electrodes, Cardiff, 1973, paper 17.
13. D. L. Manning, J. R. Stokely and D. W. Magouyrk, *Anal. Chem.* **46**, 1116 (1974).
14. L. A. R. Pioda, V. Stankova and W. Simon, *Anal. Lett.* **2**, 665 (1969).
15. U. Fiedler and J. Růžička, *Anal. Chim. Acta* **67**, 179 (1973).
16. J. Pick, K. Tóth, E. Pungor, M. Vasák and W. Simon, *Anal. Chim. Acta* **64**, 477 (1973).
17. R. P. Scholer and W. Simon, *Chimia* **24**, 372 (1970).
18. D. Ammann, M. Güggi, E. Pretsch and W. Simon, *Anal. Lett.* **8**, 709 (1975).
19. R. J. Levins, *Anal. Chem.* **43**, 1045 (1971); **44**, 1544 (1972).
20. D. Ammann, E. Pretsch and W. Simon, *Anal. Lett.* **7**, 23 (1974).
21. M. Semler and H. Adametzová, *J. Electroanal. Chem.* **56**, 155 (1974).
22. E. W. Baumann, *Anal. Chem.* **47**, 959 (1975).
23. G. Eisenman, in *Ion-Selective Electrodes*, R. A. Durst, editor, Chapter 1. N.B.S. Spec. Publ. No. 314, Washington, D.C. (1969).
24. W. Simon, W. E. Morf and P. C. Meier, *Structure and Bonding* **16**, 113 (1973).
25. W. E. Morf, D. Ammann, E. Pretsch and W. Simon, *Pure Appl. Chem.* **36**, 421 (1973).
26. E. Eyal and G. A. Rechnitz, *Anal. Chem.* **43**, 1090 (1971).
27. G. A. Rechnitz and E. Eyal, *Anal. Chem.* **44**, 370 (1972).
28. W. E. Morf, E. Lindner and W. Simon, *Anal. Chem.* **47**, 1596 (1975).
29. W. E. Morf, G. Kahr and W. Simon, *Anal. Lett.* **7**, 9 (1974).
30. W. E. Morf, D. Ammann and W. Simon, *Chimia* **28**, 65 (1974).
31. J. N. Butler, in *Ion-Selective Electrodes*, as Ref. 23, Chapter 5.
32. J. W. Ross, in *Ion-Selective Electrodes*, as Ref. 23, Chapter 2.
33. D. Langmuir and R. L. Jacobson, *Env. Sci. Technol.* **10**, 834 (1970).
34. S. S. Potterton and W. D. Shults, *Anal. Lett.* **1**, 11 (1967).
35. K. Srinivasan and G. A. Rechnitz, *Anal. Chem.* **41**, 1203 (1969).
36. G. A. Rechnitz and T. M. Hseu, *Anal. Lett.* **1**, 629 (1968).
37. R. J. Baczuk and R. J. DuBois, *Anal. Chem.* **40**, 685 (1968).
38. R. F. Hirsch and J. D. Portock, *Anal. Lett.* **2**, 295 (1969).
39. G. J. Moody and J. D. R. Thomas, *Lab. Pract.* **23**, 475 (1974).

40. M. S. Frant and J. W. Ross, *Science, N.Y.* **167**, 987 (1970).
41. S. M. Hammond and P. A. Lambert, *J. Electronal. Chem.* **53**, 155 (1974).
42. B. Fleet and G. A. Rechnitz, *Anal. Chem.* **42**, 690 (1970).
43. T. H. Ryan and B. Fleet, *Proc. Anal. Div. Chem. Soc.* **12**, 53 (1975).
44. P. L. Markovic and J. O. Osburn, *Am. Inst. Chem. Eng. J.* **19**, 504 (1973).
45. B. Fleet, T. H. Ryan and M. J. D. Brand, *Anal. Chem.* **46**, 12 (1974).
46. L. E. Negus and T. S. Light, *Instrum. Technol.* **23** (December 1972).
47. E. W. Moore, in *Ion-Selective Electrodes*, as Ref. 23, Chapter 7.
48. Orion Research Inc., *Newsletter* **3**, 33 (1971).
49. A. K. Covington and R. A. Robinson, *Anal. Chim. Acta* **78**, 219 (1975).
50. Orion Research Inc., *Newsletter* **2**, 11 (1970).
51. A. Hulanicki and M. Trojanowicz, *Anal. Chim. Acta* **68**, 155 (1974).
52. P. J. Muldoon and B. J. Liska, *J. Dairy Sci.* **52**, 460 (1969).
53. Orion Research Inc., Applications Bulletin No. 8, 1970.
54. T. P. Hadjiioannou and D. S. Papastathopoulos, *Talanta* **17**, 399 (1970).
55. R. C. Knupp, *Ceram. Bull.* **49**, 773 (1970).
56. K. L. Cheng and K. Cheng, *Mikrochim. Acta* 385 (1974).
57. Orion Research Inc., *Analytical Methods Guide*, 7th edn (1975).
58. P. J. Milham, A. S. Awad, R. E. Paull and J. H. Bull, *Analyst* **95**, 751 (1970).
59. P. J. Milham, *Analyst* **95**, 758 (1970).
60. N. G. Bunton and N. T. Crosby, *Water Treat. Exam.* **18**, 338 (1969).
61. L. J. Forney and J. F. McCoy, *Analyst* **100**, 157 (1975).
62. S. E. Manahan, *Anal. Chem.* **42**, 128 (1970).
63. J. Mertens and D. L. Massart, *Bull. Soc. Chim. Belg.* **82**, 179 (1973).
64. J. L. Walker, *Anal. Chem.* **43**, No. 3, 89A (1971).
65. L. Mela, E. Láncos, K. Ikrényi-Behany, A. G. B. Kovách, J. Pick and E. Pungor, Paper 11-8 presented at *Euroanalysis II*, Budapest, 1975.
66. I. Sekerka and J. F. Lechner, *Anal. Lett.* **7**, 463 (1974).
67. H. B. Herman and G. A. Rechnitz, *Anal. Lett.* **8**, 147 (1975).
68. R. De Wolfs, G. Broddin, H. Clysters and H. Deelstra, *Z. Anal. Chem.* **275**, 337 (1975).
69. J. Mertens, P. Van den Winkel and D. L. Massart, *Bull. Soc. Chim. Belg.* **83**, 19 (1974).

CHAPTER 7

GAS-SENSING PROBES

The development in the last few years of a range of gas-sensing probes, complementary to the ion-selective electrodes already available, has enlarged the range of species which may be analysed by potentiometry. The probes sense the partial pressure of gases in solution and suffer no interference from ions; since samples usually contain few gases to which they can respond, the probes are generally more selective than all types of ion-selective electrodes. They are commonly used to determine ions such as carbonate, ammonium or sulfite which can be simply converted to gases in solution by sample pretreatment. Such pretreatment, usually addition of either an acid or an alkali, is essential in the analysis of nearly all samples.

The concept of measuring the partial pressure of a gas in a sample by determining the pH of a thin film of solution, separated from the sample by a hydrophobic gas-permeable membrane, was first set out in 1957 by Stow, Baer and Randall,[1] who constructed a carbon dioxide probe for measurements in blood. Severinghaus and Bradley[2] improved the probe; this and later work has been reviewed by Severinghaus,[3] who studied the behaviour of the probe with various membranes and internal electrolytes. Although the carbon dioxide probe, sometimes known as the 'Severinghaus electrode', passed into regular use for the clinical measurement of the partial pressure of carbon dioxide in blood, it was more than a decade later that the same principles were applied to the measurement of other acidic and basic gases. The ammonia probe was the next to be developed and quickly became one of the two or three top-selling potentiometric sensors. More recently probes to measure sulfur dioxide, nitrogen oxides and hydrogen sulfide have appeared.

Two types of gas-sensing probes have been reported, the gas-sensing membrane probes and the gas sensing probes without membrane ('air-gap electrodes'[4]); both are depicted in Fig. 7.1.

The gas-sensing membrane probes contain a glass electrode with a flat or

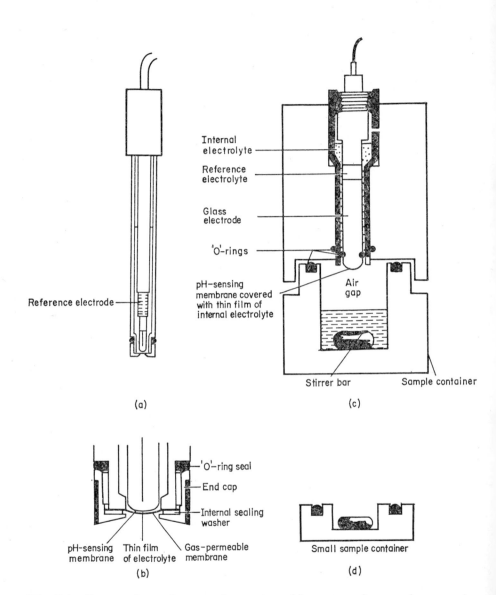

Fig. 7.1   Constructions of gas-sensing probes: (a) a gas-sensing membrane probe;
(b) cross-section of the end of the probe showing the thin film of internal electrolyte
(not to scale) ;(c) a gas-sensing probe without membrane (an 'air-gap' electrode);[37]
(d) alternative small sample container

slightly convex pH-sensitive tip. This tip is screwed down on to the membrane so that it sandwiches a thin film of the internal electrolyte against the membrane. When the probe is immersed in a sample, the determinand gas diffuses into the thin film through the air in the membrane pores or through the membrane material itself (depending on whether the membrane is microporous or homogeneous) until the partial pressures of the gas on the two sides of the membrane become equal. This equilibrium partial pressure of determinand establishes a characteristic pH in the thin film, by reaction with the internal electrolyte, and this pH is measured by means of the glass electrode and a reference electrode (usually a silver/silver halide electrode).

In the gas-sensing probes without membranes, the membrane is replaced by an air-gap of several millimetres and the probe assembly is suspended above the sample in a special sealed container.[4] The determinand gas diffuses into the thin film of electrolyte on the glass electrode tip through the air-gap. The thin film, which is usually applied to the electrode surface with a sponge before every measurement, is stabilized by addition of a wetting agent to the electrolyte. Theoretically the two types of probe are virtually identical.

So far only the gas-sensing membrane probes have been independently assessed and generally used; thus, the theory and applications sections of this chapter are devoted almost entirely to them. Some of the data also apply to the 'air-gap electrodes'. The advantages and disadvantages of the two types of probe have been argued previously;[4,5] in summary, the gas-sensing membrane probes are generally easier to use but the 'air-gap electrodes' are more satisfactory for the analysis of samples which coat or wet the membranes of particular probes.

THEORY

Gas-sensing probes are complete electrochemical cells and incorporate both an ion-selective electrode and a reference electrode within the sensor. The principle upon which they work may be stated, in a more general form than previously, as follows. When a sample is presented to a probe, determinand gas diffuses from the sample into the thin film of electrolyte on the surface of the ion-selective electrode (nearly always a pH electrode) until the partial pressures of the gas in the sample and in the thin film are equal. The determinand comes to equilibrium with the components of the thin film, and the activity of a component of the thin film is determined with the ion-selective electrode and reference electrode; this activity is related to the partial pressure or concentration of gas in the sample.

For the ammonia, carbon dioxide, sulfur dioxide and nitrogen oxide probes, the ion-selective electrode used is a glass pH electrode, and the reactions involved in the thin film are protonation equilibria; thus, the direct measurement made when a probe is used is of the pH of the thin film; types of probe incorporating ion-selective electrodes other than the glass electrode are described by Ross, Riseman and Krueger.[6] The typical components of several gas-sensing membrane probes are given in Table 7.1.

TABLE 7.1
Typical components of gas-sensing probes

Determinand	Internal electrolyte	Internal ion-selective electrode	Internal reference electrode	Membrane	Ref.
NH_3	0.1–0.01 M NH_4Cl satd with AgCl	pH electrode	Ag/AgCl	microporous PTFE	5, 6
CO_2	0.01–0.005 M $NaHCO_3$ + 0.02–0.1 M NaCl satd with AgCl	pH electrode	Ag/AgCl	microporous PTFE	6, 12
NO_x	0.1–0.02 M $NaNO_2$ + dil. NaCl or KBr satd with AgCl or AgBr	pH electrode	Ag/AgCl or Ag/AgBr	microporous PTFE or polypropylene	5, 6
SO_2	0.1–0.01 M $K_2S_2O_5$ or $NaHSO_3$ + NaCl	pH electrode	Ag/AgCl	silicone rubber or microporous PTFE	5, 6
H_2S	Citrate buffer (pH 5)	S^{2-} electrode	no information available		6

The electrochemical cell which comprises a probe may be written as follows, with an ammonia probe[5] being used as an example:

Cell 7.1

$$
\text{Ag, AgCl} \left|
\begin{array}{c} NH_4Cl\ (0.1\ M) \\ AgCl\ (s) \\ \text{\small Bulk internal} \\ \text{\small electrolyte} \end{array} \right|
\begin{array}{c} NH_4Cl\ (0.1\ M) \\ NH_3\ (aq) \\ AgCl\ (s) \\ \text{\small Thin film} \\ \text{\small Liquid junction} \end{array}
\left| \text{Glass} \right|
\begin{array}{c} KH_2PO_4\ (0.175\ M) \\ Na_2HPO_4\ (0.075\ M) \\ KCl\ (0.08\ M) \\ AgCl\ (s) \\ \text{Gelling agent} \\ \text{\small Glass electrode} \\ \text{\small internal electrolyte} \end{array}
\right| \text{Ag, AgCl}
$$

The potential of this cell, E, may be written as[7]

$$E = E°_{glass} + \frac{2.3RT}{F} \log_{10} a_{H^+} - E°_{Ag/AgCl} + \frac{2.3RT}{F} \log_{10} a_{Cl^-} \quad (7.1)$$

$$= E' + \frac{2.3RT}{F} \log_{10} a_{H^+} \quad (7.2)$$

where a_{H^+} is the activity of hydrogen ions in the thin film and a_{Cl^-} is constant. But

$$K_1 = \frac{a_{NH_4^+}}{a_{H^+} P_{NH_3}} \quad (7.3)$$

Thus,

$$E = E'' - \frac{2.3RT}{F} \log_{10} P_{NH_3} \quad (7.4)$$

where P_{NH_3} is the partial pressure of ammonia in the sample or thin film (since these are equal at equilibrium). By Henry's Law

$$P_{NH_3} = H[NH_3] \qquad (7.5)$$

where H is the Henry's Law constant. Hence, the cell potential is related to the concentration of ammonia or ammonium ion (after pretreatment) in samples by the Nernstian equation:

$$E = E''' - \frac{2.3RT}{F} \log_{10} [NH_3] \qquad (7.6)$$

Analogous equations may be written for other probes, with the negative sign replaced by a positive sign when the determinand is an acidic gas.

Response range

The upper Nernstian limit is the point at which the assumption made in Eq. (7.4), that $a_{NH_4^+}$ remains constant, ceases to be valid. This occurs when the partial pressure of determinand (e.g. ammonia) in the thin film is so high that sufficient of it hydrolyses to produce a significant increase in the activity of the ionic form of the determinand (e.g. ammonium ions) present in the internal electrolyte of the cell.

The causes of departure from Nernstian behaviour at low partial pressures of determinand vary from probe to probe,[5,6] and are sometimes difficult to explain theoretically. Two common causes are the presence in reagents of a background concentration of the determinand, which is difficult to remove, or interference caused either by external interferents or by components of the internal electrolyte. Another important factor is the rate of interchange of the internal electrolyte across the boundary of the thin film with the bulk of the electrolyte in the probe. In principle as this rate decreases the limit of detection which may be achieved also decreases. With membrane probes the rate is fairly slow and may be decreased by increasing the tension of the internal ion-selective electrode on the membrane; however too great a tension may result in membrane breakage or in drift due to osmotic problems.

The relationship between response range and the concentration of the internal electrolyte has been derived for the sulfur dioxide probe[6] and the carbon dioxide probe:[3] in neither case was the theory completely successful in describing the observed response.

Response time

A useful steady state model to describe the response time characteristics of gas-sensing membrane probes has been proposed by Ross et al.[6] A complex equation was derived for the time, t, taken by a probe to achieve a given fraction $(1 - \varepsilon)$ of the final potential change after a change of determinand concentration from C_1 to C_2. This equation was soluble, if certain assumptions were made, yielding the expression:

$$t = \frac{lm}{Dk}\left(1 + \frac{dC_B}{dC}\right) \ln \frac{\Delta C}{\varepsilon C_2} \tag{7.7}$$

where l = thickness of the thin film;
$\quad\quad m$ = thickness of the membrane;
$\quad\quad D$ = diffusion coefficient of the determinand in the membrane;
$\quad\quad k$ = partition coefficient of the determinand between the sample or thin film and the membrane;
$\quad\quad C_B$ = sum of the concentrations of all forms of the determinand in the internal electrolyte other than the gaseous form;
$\quad\quad C$ = concentration of the gaseous form of the determinand in the internal electrolyte;
$\quad\quad \Delta C = C_2 - C_1$;
$\quad\quad \varepsilon$ = fractional approach to equilibrium = $\left|\dfrac{C_2 - C}{C_2}\right|$.

In addition to predicting the effect of membrane properties and the thickness of the thin film on the response time, the equation also predicts that the response time is dependent on the direction of concentration change. For concentration changes from low to high ($C_2 \gg C_1$), the equation predicts that the response time will be almost independent both of the magnitude of the concentration change and of the magnitudes of C_1 or C_2: this prediction is supported by the experimental work of Bailey and Riley,[5] who showed that the response times of an ammonia probe for decadic concentration increases were indeed independent of the final concentration at ammonia concentrations above 10^{-4} M. Equation (7.7) also suggests that the response times for concentration increases will be much shorter than for concentration decreases, as is observed in practice: for the concentration change from 10^{-5} to 10^{-1} M determinand, the response time, t_{99} ($\varepsilon = 0.01$), is calculated[6] to be 13 times less than for the reverse change.

Membranes and osmotic effects

In gas-sensing membrane probes two types of membrane are commonly used, microporous and homogeneous. The microporous membranes consist of an inert, hydrophobic matrix (typically made of PTFE) containing tiny pores normally filled with air; the determinand gas diffuses from the sample to the thin film through the air in the membrane pores. In contrast, the homogeneous membranes are solid films of plastics such as silicone rubber; in these membranes the determinand gas diffuses through the membrane material. The value of Dk (Eq. 7.7) for any gas is substantially lower in the homogeneous membranes than in the microporous membranes owing to the much larger values for D in air.[6]

The 'air-gap electrodes' may be fitted into the picture by considering them as probes with thick air membranes.

Gas-sensing probes are constructed such that a membrane separates a thin film of internal electrolyte from the sample. If the total concentration of dissolved species differs on the two sides of the membrane, an osmotic pressure difference results and water vapour diffuses across the membrane until the water activity is the same on each side. An osmotic pressure difference is also produced if the temperatures of the thin film and sample differ; this difference can lead to very much larger rates of diffusion of water than those produced by differences in the total concentration of dissolved species.[6]

If diffusion of water vapour across the membrane does occur, the thin film will be either concentrated or diluted, depending on the direction of the diffusion. This will result in an osmotic drift in the probe potential at a rate related to the rate of transfer of the water vapour. At very low rates of diffusion, the interchange between the thin film and the bulk of internal electrolyte may make the concentration or dilution effects negligible. The magnitude of the drift caused by osmotic pressure differences is thus clearly minimized if the value of Dk for water vapour in the membrane material is a minimum. However, as the value of Dk for water vapour decreases with decreasing membrane permeability, so the value of Dk for the determinand gas will decrease and the response time lengthens. Such an increase in response time may well limit the application of the probe and effectively increase the limit of detection. So a compromise must be reached in an attempt to minimize both the osmotic drift and the response time. In practice, for many probes, such as the ammonia probe, a choice of membrane type does not exist because the values of Dk are too small in all but the microporous membranes. However, for the sulfur dioxide probe, for example, the Dk value for sulfur dioxide in silicone rubber is sufficiently high for that membrane material to be used, with concomitant reduction of osmotic effects. A list of relevant Dk values is given in Table 7.2 (adapted from Ref. 6): Ross et al.[6] calculated that, to produce an analytically useful probe, the value

TABLE 7.2
Values of Dk (cm^2 s^{-1}) for various gases and diffusion media (adapted from Ref. 6)

Gas	Air (microporous membrane or air-gap)	Dimethyl silicone rubber (homogeneous membranes)	Low-density polythene (homogeneous membranes)
CO_2	1.6×10^{-1}	2.9×10^{-5}	3.9×10^{-7}
NH_3	5.3×10^{-4}	9.8×10^{-8}	
SO_2	3.7×10^{-3}	3.8×10^{-6}	
NO_2	2.2×10^{-3}	1.0×10^{-6}	
H_2S	7.7×10^{-2}	3.4×10^{-5}	
CH_3COOH	2.5×10^{-6}		
H_2O	1.3×10^{-7}	5.1×10^{-9}	

of Dk for the determinand in the probe membrane should not be less than approximately 10^{-6} cm^2 s^{-1}.

As shown in Table 7.2, the value of Dk for water vapour in microporous membranes (diffusion through air) is relatively large; hence, when probes with microporous membranes are used, it is important to ensure that the osmotic strength difference between samples and the internal electrolyte is minimized. This is achieved by careful selection of the strengths of reagents for sample pretreatment or, occasionally, by adjustment of the internal electrolyte so that the osmotic strengths of the two solutions are matched. It is also important to ensure that both sample and probe are at the same temperature. Osmotic drift is particularly noticeable if the osmotic strength of the sample is greater than that of the internal electrolyte (as in the analysis of Kjeldahl digests with the ammonia probe): if sufficient dilution of the sample is not practicable, then the internal electrolyte must be modified by addition of a suitable inert electrolyte. It cannot be satisfactory to increase arbitrarily the strength of pretreatment buffers without changing the internal electrolyte.[8] These remarks apply equally to gas-sensing membrane probes with microporous membranes and to 'air-gap electrodes'. However, for gas-sensing membrane probes with homogeneous membranes, such as the sulfur dioxide probe with a silicone rubber membrane, the osmotic drift is reduced to such an extent that it is too low for measurement.[5] In such cases no special care need be taken to match osmotic strengths or temperatures; this is particularly convenient when samples, such as glucose syrups or fruit concentrates, which have a high and unknown osmotic strength are analysed.

Selectivity

A gas-sensing probe can only suffer direct interference from dissolved gaseous species in the treated sample which either change the activity of the species sensed by the internal ion-selective electrode or interfere with the response of that electrode. If, as is usual, the ion-selective electrode is a pH electrode, the possibility of interference with the response of the electrode to hydrogen ions may be safely ignored. Thus, the only species which interfere with the probe response are gases which can diffuse through the membrane relatively fast and have an acidity comparable to or greater than that of the determinand (or basicity for probes sensing alkaline gases); the degree of interference can thus be estimated from a comparison of the values of Dk, pK and concentration for the determinand and interferent.

Although the conditions under which interference can occur are clear, it is difficult to quantify the amount of interference expected in a given set of practical conditions because, for example, of the indeterminate rate of exchange of electrolyte between the thin film and the bulk of the solution. The usual symptom of interference with the probe response is drift of the potential; hence, the concept of selectivity coefficients is not useful for gas-sensing probes. It is always more satisfactory to eliminate interferents by suitable sample pre-treatment than to try to allow for their effect on the probe potential. However possible interferents are very few for all the probes and selectivity is very seldom an important problem in their application. That this is true is a measure of their high selectivity.

PERFORMANCE

Response range

The response ranges of the gas-sensing probes, in terms of the concentrations of the determinands as gases, are given in Table 7.3. The two ammonia and nitrogen oxide membrane probes are bracketed together as their performances are similar: the two sulfur dioxide probes however, are, dissimilar, because the E.I.L. probe uses a silicone rubber (homogeneous) membrane, whereas the Orion probe has a microporous membrane.

Table 7.3
Response ranges of gas-sensing probes

Probe	Type	Upper useful limit/M	Limit of Nernstian response/M (mg l^{-1})	Limit of detection/M	pH require-ment	Ref.
NH_3	E.I.L./Orion	1	ca. 10^{-6} (0.02)	ca. 10^{-7}	>12	5, 6, 7, 9–11
	'air-gap'		10^{-4} (2)	10^{-5}	>12	4
CO_2	Radiometer		10^{-5} (0.4)	2×10^{-6}	<3.4	12
	'air-gap'	1	2×10^{-3} (90)	10^{-4}	<3.4	13
SO_2	E.I.L.	5×10^{-2}	5×10^{-5} (3)	5×10^{-6}	<0.7	5
	Orion	10^{-2}	5×10^{-6} (0.3)	5×10^{-7}	<0.7	14
NO_x	Ref. 5, 16/Orion	10^{-2}	2×10^{-6}	10^{-7}	<2	5, 6, 15, 16
H_2S	Orion	10^{-2}		10^{-8}	<5	6

The upper useful limits of the probes are set by the onset of potential drift. Typically, if a sample is above the quoted limit for a particular probe, the probe drifts continually to read higher and higher concentrations: this limit is usually somewhat lower than that predicted by theory as the upper limit of Nernstian response.

The parameters responsible for the values of the lower limits of detection of the probes vary from probe to probe. The practical limit of detection of the ammonia probe is set by the purity of the water available to prepare the standards and reagents;[7,9,11] it is very difficult to prepare water containing less than 0.01 mg N per litre (about 10^{-6} M). Long response times also limit the use of the probe below 0.05 mg N per litre.

In order to achieve the above limit of detection for the carbon dioxide probe, Midgley[12] rigorously excluded atmospheric carbon dioxide from his apparatus and degassed the acidic reagent: the parameters limiting the probe response were not identified. However it is possible that the observed limit is set by the concentration of carbon dioxide in the thin film when equilibrium has been achieved between the passage of the gas through the thin film through the membrane, the generation of the gas by hydrolysis of the internal electrolyte and the interchange of that electrolyte across the boundary of the thin film. The analogous 'air-gap electrode' is curiously insensitive for no apparent reason; although the acidic reagent was not degassed to remove atmospheric carbon dioxide, this would only introduce a small fraction of the carbon dioxide required to explain the very high limit of detection. The detection limit of the Orion sulfur dioxide probe is probably due to the same equilibria as suggested for the carbon dioxide probe. For the E.I.L. sulfur dioxide probe, the somewhat higher detection limit is observed because an acidic preservative is added to the internal electrolyte to prolong its life.

The species sensed by the nitrogen oxide probes is the mixture of nitrogen oxides generated when a nitrite solution is acidified; hence, nitrite is considered as the determinand. The measured limits of detection of the nitrogen oxide probes may be attributed to interference by atmospheric carbon dioxide, both in the water used to make the standards and in the acidic reagent. If carbon dioxide were rigorously excluded, it seems probable that the measured limits of detection could be decreased.

The pH values for samples to be measured with the probes are listed in Table 7.3; the limiting values given are those at which virtually all of the determinand is present in the samples as dissolved gas. These values may be simply calculated from the stability constants of the various species.

Selectivity

A summary of published selectivity data is given in Table 7.4. The compounds which interfere are generally volatile acids or bases and follow the behaviour predicted theoretically. Several authors have shown that ions in treated samples do not interfere.

Once a gas-sensing probe has been exposed to a sufficiently large concentration of interferent to give severe interference, the response of the probe becomes very sluggish; renewal of the thin film is the quickest way to restore performance.

The selectivity of the 'air-gap electrodes' is probably similar to that of the gas-sensing probes with microporous membranes.

Response time

The response times of E.I.L. ammonia and sulfur dioxide probes and a nitrogen oxide probe have been reported by Bailey and Riley:[5] these results are given in Table 7.5. The definition of response time used is that provisionally recommended by IUPAC (see Chapter 1). It may be seen that, as theoretically predicted, the response times for increases in concentration are considerably shorter than for decreases. The response times of an ammonia probe for

TABLE 7.4
Selectivity data for gas-sensing membrane probes

Probe	Interferent	Concn of interferent	Concn of determinand	Apparent increase in determinand concn	Ref.
NH_3	hydrazine	$4\,mg\,l^{-1}$	$1\,mg\,l^{-1}$	$0.06\,mg\,l^{-1}$	7
	cyclohexylamine	$1\,mg\,l^{-1}$	$1\,mg\,l^{-1}$	$0.03\,mg\,l^{-1}$	
	morpholine	$10\,mg\,l^{-1}$	$1\,mg\,l^{-1}$	$0.03\,mg\,l^{-1}$	
	octadecylamine	$0.4\,mg\,l^{-1}$	$1\,mg\,l^{-1}$	$0.14\,mg\,l^{-1}$	
	methanolamine	$3.4\,mg\,l^{-1}$	$0.5\,mg\,l^{-1}$	$0.15\,mg\,l^{-1}$	11
	methylamine	$2.1\,mg\,l^{-1}$	$1.2\,mg\,l^{-1}$	$0.66\,mg\,l^{-1}$	17
	ethylamine	$3.2\,mg\,l^{-1}$	$1.2\,mg\,l^{-1}$	$0.60\,mg\,l^{-1}$	
	(numerous other organic nitrogen compounds have been found not to interfere[11,17])				
CO_2	sulfite	$100\,mg\,l^{-1}$	$10\,mg\,l^{-1}$	$4\,mg\,l^{-1}$	12
	hydrochloric acid	$>0.05\,M$	$ca.\ 3 \times 10^{-4}\,M$	positive drift	18
SO_2	(E.I.L.) acetic acid	$10^{-1}\,M$	$10^{-3}\,M$	$1 \times 10^{-4}\,M$	5
	(Orion) acetic acid	$5 \times 10^{-3}\,M$	$10^{-3}\,M$	$1 \times 10^{-4}\,M$	
	(Orion) hydrofluoric acid	$3 \times 10^{-3}\,M$	$10^{-3}\,M$	$1 \times 10^{-4}\,M$	
	(Orion) hydrochloric acid	$>1\,M$ tolerable	$10^{-3}\,M$		
NO_x	(Ref. 5) carbon dioxide	$10^{-3}\,M$	$10^{-5}\,M$	$2 \times 10^{-6}\,M$	5
	(Orion) carbon dioxide	$3 \times 10^{-2}\,M$	$10^{-3}\,M$	$1 \times 10^{-4}\,M$	

successive concentration decreases have also been measured by Gilbert and Clay.[10] The response time increased steadily from less than 1 min at 10^{-4} M NH_3 to 7 min at 10^{-6} M and 145 min at 1.6×10^{-8} M: all the samples except the first were buffered with respect to ammonia (i.e. they were ammonium chloride solutions adjusted to a range of different pH values). These results are in satisfactory agreement with those of other authors.[9,11] The ammonia-sensing 'air-gap electrode' is reported to reach a steady state within 1–2 min for samples in the concentration range 3×10^{-5}–6×10^{-4} M NH_4^+.[4]

TABLE 7.5
Results of duplicate determinations of probe response times

Final concn/M	Ammonia	Sulfur dioxide	Nitrogen oxide
	Time/s		
	Decadic concentration increases		
10^{-5}	105, 110		370, 480
10^{-4}	32, 33	425, 335	95, 92
10^{-3}	31, 31	36, 36	34, 38
10^{-2}	31, 30	32, 27	
	Decadic concentration decreases		
10^{-5}	120, 70		460, 410
10^{-4}	66, 34	155, 155	125, 66
10^{-3}	32, 30	68, 42	

The Radiometer carbon dioxide probe studied by Midgley was used in a closed flow system.[12] The response time of the system, which included approximately 2 min wash-out time for the flow cell, was 3–7 min to reach equilibrium for a tenfold concentration increase for carbon dioxide concentrations above 1 mg l^{-1} (about 2×10^{-5} M). The response times for concentration decreases were generally approximately double those for increases and extended to 30 min for concentrations less than 1 mg l^{-1}. Růžička and Hansen[4] quote the 95% response time for the carbon dioxide-sensing 'air-gap electrode' as 30 s for carbon dioxide concentrations in the range $3 \times 10^{-3} - 5 \times 10^{-2}$ M.

Temperature effects

Changes in the temperature of a probe and samples affect the probe response in several ways. When the probe and samples are at the same temperature and temperature changes are sufficiently slow for the system to remain virtually at equilibrium, the slopes of the probes vary approximately as predicted by the Nernst factor ($2.3RT/F$) and the variation of the standard potentials may be expressed by temperature coefficients. Measurements of temperature coefficients have been made by warming up the probes and samples simultaneously in a thermostatted cabinet; the results are given in Table 7.6. These values will be the algebraic sum of the temperature coefficients of all the individual components of Cell (7.1).

TABLE 7.6
Temperature coefficients of gas-sensing membrane probes

Probe	Temperature coefficient/mV K^{-1}	Temperature range/K	Ref.
NH_3 (E.I.L.)	1.5	16–29	7
NH_3 (E.I.L.)	1.3	28–35	5
SO_2 (E.I.L.)	0.5	26–43	5
NO_x (Ref. 5)	0.2	25–31	5
NO_x (Ref. 5)	0.0	31–45	5
CO_2 (Radiometer)	1.0	15–35	12

However, the temperature effect which frequently predominates, and is more often than not the cause of drift, is osmotic transfer of water vapour caused by a temperature gradient across the membrane. If the sample temperature changes, the temperature of the thin film is slow to follow that change because of the insulating effect of the membrane and the large thermal mass of the pH electrode in intimate contact with the film. Hence, the partial pressures of water vapour on the two sides of the membrane differ and water vapour transport occurs, particularly through microporous membranes. Eventually, after the inside of the probe has reached the new sample temperature, the process will reverse and water vapour will move in the opposite direction until the partial pressures are again equal.[5] The result of this complex behaviour is a curve such as that shown in Fig. 7.2 for the ammonia probe (curve A). If the amount of water vapour transferred is very small, as occurs if the membrane is homogeneous, the characteristic dip in the curve is not observed. The potential change is then almost entirely due to the change in the standard potential; thus, a smooth curve is obtained with a sulfur dioxide probe with a homogeneous silicone rubber membrane (curve B). The behaviour shown in curve A has been observed by other authors.[10,12] Hence, it is particularly important when gas-sensing probes with microporous (or air) membranes are used to ensure that the temperatures of both the samples and the probe are adequately stable and identical. It is even advisable not to use the probes in direct sunlight for this reason.

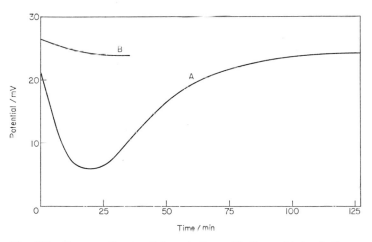

Fig. 7.2   Curves of potential variation with time obtained after a temperature change for the ammonia and sulfur dioxide probes.[5] A, Ammonia probe: $[NH_3] = 6 \times 10^{-4}$ M; 31.3–34.6 °C. B, Sulfur dioxide probe: $[SO_2] = 10^{-3}$ M; 31.0–36.7 °C

Osmotic effects

Osmosis occurs, as described earlier in the chapter, when the partial pressures of water vapour on the two sides of the membrane differ. There are two principal situations in which this may arise. The first, when there is a temperature gradient across the membrane, has already been described. The other situation in which osmosis occurs is when the total concentration of dissolved species is different on either side of the membrane. Thus, attempted measurements of ammonia directly in Kjeldahl digests or sea-water with an unmodified ammonia probe are subject to gross drift.

The drift of an ammonia probe with a microporous membrane in a 5×10^{-4} M solution of ammonia in 3 M sodium chloride amounted to approximately -17 mV min^{-1}, which is greatly in excess of what can be tolerated. In contrast, a sulfur dioxide probe with a homogeneous membrane showed no measurable drift in a 10^{-3} M solution of sulfur dioxide in 3 M sodium chloride.[5]

For the analysis of samples containing a high total concentration of dissolved species, osmotic drift may be prevented either by sufficient dilution of the sample or, when that is not realistic or possible, by dosing the internal electrolyte of the probe to match the samples. Further mention of this subject is made in the section on Applications.

Lifetime

The lifetime of gas-sensing membrane probes may be assessed by the necessary frequency of re-assembly, which is usually a measure of the length of time the membrane lasts. The other components in the probes are relatively durable.

Beckett and Wilson[11] have used an ammonia-sensing membrane probe for the analysis of fresh waters; they found that a membrane lasted approximately 3 months on average. An ammonia probe used by Midgley and Torrance[7,19] showed no deterioration in performance over 4 months continuous use; similarly, Mertens et al.[20] found that the lifetime was 2 months. In the analysis of ammonia in a wide variety of samples, including trade wastes and sewage, Evans and Partridge[21] adopted the procedure of changing the probe membrane every month. However, Proelss and Wright[22] were only able to use membranes for 1–2 weeks when using an ammonia probe for the analysis of ammonia in treated blood samples.

Midgley[12] has reported the continuous use of a single carbon dioxide probe assembly for 10 weeks with no deterioration of the performance.

Membrane failure, which is the usual limiting factor on the lifetime of a particular probe assembly, usually occurs because the membrane is broken or becomes coated, or, in the case of microporous membranes, because the hydrophobicity is destroyed. Membrane breakage may usually be attributed to misuse, such as allowing the probe to be hit by a stirrer bar or puncturing the membrane by screwing the glass electrode too far down. However, the membrane life will also be shortened if the probe is used in samples containing very finely divided particles or gelatinous material, which can block the pores or surface of the membrane. Such blockage can usually be avoided by appropriate sample pretreatment to minimize the amount of precipitated material or gels. Occasionally it is either not convenient or impossible to remove or dissolve all the harmful material and the probe tip must be washed after every set of analyses. Thus, for example, in the analysis of sulfur dioxide in sausage meat with a sulfur dioxide probe, the membrane becomes coated with fat and protein and is cleaned *in situ* by periodic soaking in sodium hydroxide solution. The presence of wetting agents in samples shortens the life of microporous membranes[6] by slowly destroying the hydrophobicity of the membrane and making it permeable to water; this problem does not affect probes with homogeneous membranes or 'air-gap electrodes'. Membranes which have failed may often be given a second lease of life by carefully washing them and letting them dry out completely.

The other components of gas-sensing probes which will also fail eventually include the glass electrode and reference electrode and internal electrolyte, but several years' life may be expected for all of these. The glass electrode, which is in constant contact with poorly buffered solutions at close to neutral pH values (almost minimum conditions for prolonged successful operation), will eventually become sluggish in response and may be reactivated by cleaning with an organic solvent, such as isopropanol, to remove any grease, followed by an overnight soak in dilute hydrochloric acid. The reference electrode may usually be replated to give a new silver halide coating or abraded to expose a new layer. For some probes, such as the sulfur dioxide probe, deterioration of the internal electrolyte may also be partly responsible for the required frequency of re-assembly.

APPLICATIONS

Ammonia probe

The principal application of the ammonia probe is in the analysis of fresh water, effluents and treated and untreated sewage. The most thorough studies of the ammonia probe for this purpose are those of Beckett and Wilson[11] and Evans and Partridge.[21] Both groups of workers used, as alkaline buffer, a solution of sodium hydroxide with EDTA added to prevent precipitation of metal hydroxides which might coat the membrane. Satisfactory concentrations of the components are 1 M sodium hydroxide + 0.1 M EDTA; one volume of buffer is to be added to ten volumes of sample. On the assumption that samples have a relatively low total concentration of dissolved species (<0.02 M), this buffer ensures that the osmotic pressures of both the samples and the thin film are closely similar and that the pH of the samples is raised to greater than 12.

Beckett and Wilson[11] compared methods for the analysis of ammonia in treated and untreated water from the River Thames based on the ammonia probe, an automated indophenol blue procedure and a distillation–Nesslerization procedure. They showed that there was a 'lack of important bias' in the results from the probe and that the relative total standard deviation, of the results on standard solutions, varied from 10 % in the range 0.05–0.1 mg N per litre to 3 % between 0.4 and 4 mg N per litre. A wide range of samples, from crude sewage to potable water, were analysed for both total ammoniacal and albuminoid nitrogen by Evans and Partridge,[21] who compared the results with those from an indophenol blue method and a Nessler method. Samples containing chlorine, such as swimming-pool water, were treated with sodium thiosulfate solution before alkaline treatment to prevent interference from chloramines. In a series of experiments on spiked samples, the average recovery was excellent (9.9 mg l^{-1} for 10 mg l^{-1} added, 1.00 mg l^{-1} for 1.00 mg l^{-1} and 0.10 mg l^{-1} for 0.10 mg l^{-1}) with satisfactory coefficients of variation (3.2 %, 3.3 % and 16.0 %, respectively). The precision of determination of ammoniacal nitrogen was found to be better than 4 % for concentrations greater than 0.4 mg N per litre and within 0.105 mg N per litre for less than 0.4 mg N per litre. The determinations of albuminoid nitrogen using the probe were also in satisfactory agreement with the other methods except when the ratio of free to albuminoid nitrogen in the samples was large.

Measurements in surface waters, sewage and saline waters have also been made by Thomas and Booth,[9] who showed that the accuracy of the ammonia probe method was equal to that of an automated indophenol blue method but the probe method was slightly more precise. Similar results have been obtained by other workers.[10,17] Sekerka and Lechner[23] have automated the analysis of samples with the ammonia probe. Samples in the concentration range 0.1–100 mg NH_4^+ per litre were analysed with standard deviations from 11% down to 2%: the sampling rate of 20 per hour for concentrations greater than 10 mg NH_4^+ per litre and only 10 per hour for lower concentrations was presumably limited by the wash-out time of the system or the response times of the other sensors in the system. Other workers[5,24] have analysed ammonia samples in concentration ranges from 1 to 15 mg l^{-1} at the rate of 60 per hour in continuous flow systems: an example of a recorder trace from such a system is given in Chapter 9 (p. 209).

The determination of ammonia, in condensed steam and boiler feed-water, with an ammonia probe has been investigated by Midgley and Torrance[7,19] and Mertens et al.[20] Results from an ammonia probe method were compared[7] with results from a method using an ammonium-selective glass electrode; they differed by a maximum of 6% and an average of 0.3%. In some waters the results given by the glass electrode had to be corrected for interference from sodium ions. When used in a continuous flow system[19] with automatic standardization every 12 h, ammonia concentrations in the range 0.1–1.0 mg NH_3 per litre could be determined with a relative standard deviation of at most 10%. Mertens et al.[20] also used a continuous flow system in their comparison of the ammonia probe with the ammonium electrode based on neutral carriers; both potentiometric systems agreed well with spectrophotometric methods and gave good (about 2%) accuracy and reproducibility (mean deviation from the mean of 1–2%). In practice, in boiler-feed waters and steam condensates in which the concentration of sodium ions is low, the ammonium-selective glass electrode is preferred because it requires virtually no maintenance and is cheaper; the greater selectivity of the probe which makes it superior for the analysis of other samples is here unimportant.

The high selectivity of the ammonia probe, in comparison with the nitrate electrode, has led several workers to try to use it to measure nitrate after reduction to ammonia: thus, waters containing high concentrations of chloride ions, bicarbonate ions, iodide ions and other species which interfere with the nitrate electrode, may be analysed without interference with the sensor. An interesting and useful study of this method for nitrate has been made by Mertens, Van den Winkel and Massart.[25] They concluded that the most satisfactory reductant was Devarda's alloy powder bonded into beads with polystyrene or PVC; the reduction occurred most rapidly at pH 13. The measurement was adapted for use in the continuous-flow analysis system shown in Fig. 7.3, and gave a linear calibration graph in the range 1–50 mg NO_3^- per litre with a sampling rate of 20 per hour. The reduction efficiency was 100% for concentrations in this range with a freshly prepared reduction column; after 5 days the reduction efficiency had dropped to 80% but the calibration graph was still linear. For samples containing 3 and 30 mg NO_3^- per litre the mean measured values were 3.01 and 29.6 mg l^{-1}, respectively, with relative standard deviations of 2.5% and 1.85%.

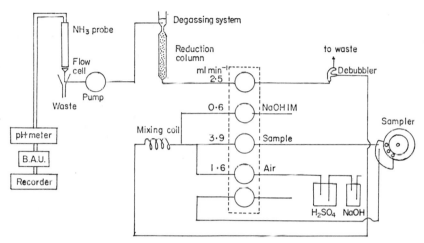

Fig. 7.3   Flow diagram of an automatic system for the determination of nitrate with an ammonia probe[25]

Another major application of the ammonia probe is in the analysis of ammonia in Kjeldahl digests: use of the ammonia probe avoids the necessity to distil the ammonia out of the digest for measurement and thus considerably shortens the analysis time. The digests may be analysed at the rate of 60 per hour with a continuous-flow system such as that described by Buckee.[24] The osmotic strength of Kjeldahl digests is always very high; hence, it is usual to dilute the samples before making them alkaline and presenting them to the probe. Even so, the total concentration of dissolved solids in the samples (after treatment with the alkaline buffer) is usually higher than in the internal electrolyte and it is essential to modify the internal electrolyte to prevent osmotic drift. This may be done most easily by addition of sodium sulfate to the internal electrolyte to match it to the samples; alternatively, it is satisfactory, when working well away from the limit of detection of the probe, to increase the concentration of ammonium chloride.[24] It is also advisable to control the temperature of samples, since the reaction between the alkaline buffer and the acidic digest generates a significant amount of heat. Catalysts in the digest, such as mercury or copper, must be removed by addition of iodide[26] or EDTA, respectively, to the buffer to prevent interference. Reports have appeared on the use of the ammonia probe for the analysis of total nitrogen in the Kjeldahl digests of soil,[26] meat products,[27,28] cereals,[28] malt, wort and beer.[24] The use of a continuous-flow system gives the most satisfactory results, since the conditions of measurement are standardized and the temperature closely controlled. Buckee,[24] using such a system, tested the accuracy by performing replicate analyses of pure acetanilide and glycine, which contain 10.36% w/w and 18.66% w/w nitrogen, respectively. The average of ten estimations of each gave 10.33% (standard deviation, 0.04%) and 18.64% (standard deviation, 0.02%), respectively.

The determination of ammonia in blood with an ammonia probe has been described by several authors.[29-31] Park and Fenton[29] used a continuous-flow system to analyse plasma samples at the rate of 20 per hour; the samples were made alkaline by the addition of 0.1 M potassium carbonate solution at a ratio of 1 volume of sample to 0.8 volume of buffer. Excellent recoveries were obtained from replicate analyses of spiked samples of pooled plasma. The probe method was described as both simple and reliable. Proelss and Wright[22] found it necessary to precipitate out the proteins in the sample by addition of perchloric acid; the samples were then centrifuged and the supernatant liquid was separated for analysis. They also modified the internal electrolyte of the probe to match the osmotic strength to that of the samples. Increase in the apparent ammonia concentration due to acceleration of ammonia-generating reactions in the alkaline conditions of measurement was also discussed and can produce significant errors: these errors were minimized by cooling the samples. Using a manual (unautomated) method, the authors found that the coefficients of variation of the results ranged from 3.5% to 4.8%, depending on the concentration range. The limit of detection of the method was 20 μg NH_3 nitrogen per litre of blood. Ammonia in blood has also been analysed satisfactorily with the 'air-gap electrode'.[4]

The ammonia probe has also been used for pharmaceutical analysis. Van-Vlasselaer and Crucke[31] analysed N-unsubstituted carbamates such as meprobamate by decomposing them in basic alcoholic solution, hydrolysing the resultant cyanate to ammonium ions in acidic solution, and finally making the solution alkaline again to determine the ammonium concentration. The method is fast compared with conventional methods. Determinations of ammonia in airborne particulates[32] and in tobacco smoke[33] have been reported. The ammonia, or the particles on which it is absorbed, is scrubbed or filtered out from the gas and measured conventionally with a probe.

The several enzymatic methods of analysis which have been developed based on an ammonia probe are discussed in the following chapter.

In general, the principles involved in the development of methods with the ammonia probe are the same as for other gas-sensing probes. Samples should be stirred, the temperature of both the probe and samples should be the same and the pH should be correctly adjusted. No separate reference electrode is needed, as the probes are complete electrochemical cells incorporating their own reference electrode.

Carbon dioxide probe

The carbon dioxide probe was developed for the measurement of the partial pressure of carbon dioxide in blood and is now widely used for that purpose.[3] Only more recently have other applications been found. A miniaturized version[34] of the probe for biological research has been reported.

Midgley[12] investigated the feasibility of using a carbon dioxide probe for the continuous analysis of power station feed-waters and condensed steam, but the limit of detection of the probe was found to be too high under normal operating conditions of the boilers. It was concluded that the probe may be useful for the occasional monitoring of abnormal conditions.

Fiedler et al.[13] report the application of the carbon dioxide-sensing 'air-gap electrode' to the determination of the total inorganic carbon (TIC) and the total organic carbon (TOC) content of waters. The total inorganic carbon content was determined after straightforward acidification of samples to generate carbon dioxide. To analyse the organic carbon content, samples were oxidized to carbonate, by boiling under pressure for 30 min with an alkaline potassium persulfate solution, and then acidified. Such a procedure was satisfactory for carbonate concentrations down to 2×10^{-3} M, but, because of the limited Nernstian range of the sensor, preconcentration was needed for more dilute samples. This involved precipitation of the carbonate in alkaline solution as lead carbonate, by addition of lead nitrate solution, and collection of the precipitate on a filter. The filter could then be placed directly into the measurement chamber of the sensor and dissolved to generate the carbon dioxide in situ. Up to 100-fold preconcentration was possible with this technique. Samples containing 1×10^{-4}–5×10^{-3} M carbonate gave excellent recoveries, after correction for the blank value corresponding to the contribution of carbon dioxide from the air.

The determination of carbonate impurities, in nickel hydroxide intended for use in storage batteries, has been reported by Almgren.[18] The best precision (0.05–0.15 mg CO_3^{2-}) was obtained by treating 50 mg samples with hydrochloric acid, performing multiple known additions of standard carbonate solution and calculating the result by means of a Gran's plot (see Chapter 9).

Sulfur dioxide probe

Reported applications of the sulfur dioxide probe are very few. However, several unreported applications are widespread; in particular, the use of the probe for the analysis of food and beverages is slowly increasing.

Sulfur may be determined in petroleum[35] by using a sulfur dioxide probe to analyse the combustion products. The fuel is burnt in a sulfur lamp apparatus and the sulfur dioxide produced is collected and complexed in a solution of tetrachloromercurate ions buffered to pH 6.9 with a phosphate buffer and containing glycerol to inhibit oxidation. The solution is later treated with sulfamic acid to remove nitrite ions and acidified, and the sulfur dioxide determined by a known addition procedure. The optimum conditions for the

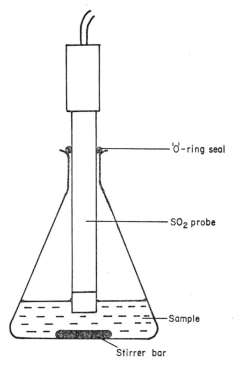

Fig. 7.4 A simple arrangement for the determination of volatile gases with gas-sensing membrane probes

collection of the sulfur dioxide were investigated and the probe results were compared with those from the standard barium sulfate gravimetric method; acceptable agreement was obtained, on taking into account the limited accuracy of the methods. The limit of the probe method was below 0.005% sulfur. The same collection medium and treatment procedure can be used for scrubbing sulfur dioxide from stack gases for measurement.

A major area of application of the sulfur dioxide probe is the analysis of food and beverages.[36] By direct acidification of liquid samples, or slurries of solid samples, to pH $\leqslant$ 0.7 with sulfuric acid, 'free' sulfur dioxide may be measured directly. Samples containing 'bound' sulfur dioxide (e.g. aldehyde-bisulfite addition compounds) may also be analysed after an alkaline pretreatment stage (pH $\geqslant$ 12) to liberate the sulfur dioxide: hence, 'total' sulfur dioxide measurements may be made. Although no reports have yet been published, such measurements of 'free' sulfur dioxide have been carried out in a wide variety of samples, including sausage seasoning, sucrose, fruit juices, fruit squashes and fruit purées. 'Total' sulfur dioxide has also been determined in beer (by standard addition techniques), wine, glucose syrups, sausages, beefburgers, fruit concentrates and fruit squashes. The whole analysis usually takes less than 10 min, which thus allows a considerable time-saving compared with traditional distillation methods. In the analysis of samples of high and undetermined osmotic strength, such as glucose syrup or fruit concentrates, the use of a probe with a homogeneous membrane avoids the need to modify the internal electrolyte.

The probe has been used in a continuous-flow system[5] to analyse standard solutions in the range 50–200 mg SO_2 per litre at the rate of 30 per hour; thus, it should also be feasible to analyse samples in such a system.

Samples may be brought to the correct pH for measurement by the addition of one volume of 2 M sulfuric acid to ten volumes of sample: stronger acid, or a different ratio, is required if the sample is strongly buffered (e.g. fruit concentrates, lime juice, etc.). Because of the high volatility of sulfur dioxide, all measurements must be carried out in a closed system. Commercially available probes fit conveniently into the top of 100 ml conical flasks; thus by fitting an 'O'-ring on to the body of the probe, and letting the 'O'-ring seal the neck of the flask under the weight of the probe, a very simple closed system is formed for routine laboratory measurements. The arrangement is shown in Fig. 7.4.

Unfortunately, the reagents available for preparing standard sulfur dioxide solutions for calibrating the probe, such as potassium metabisulfite or sodium sulfite, all have relatively low purity; usually the quoted purities are no better than 96%. Thus, for accurate work it is necessary to standardize the stock standard solution by an iodine/thiosulfate method immediately before preparing the calibration standards by dilution. Such standardization of the stock solution should be carried out each time that it is to be used, since solutions of metabisulfite or sulfite deteriorate rapidly through aerial oxidation.

Nitrogen oxide probe

The nitrogen oxide probe has, so far, been the subject of one paper,[15] in which Tabatabai describes a straightforward application of the probe to the measurement of nitrite in soil extracts and water samples. All samples were acidified with a buffer consisting of 1.34 M Na_2SO_4 + 1.00 M H_2SO_4 at a volume ratio of 10:1. The results were shown to agree well with those given by a modified Griess–Ilosvay method and, whereas the latter method is affected by sample turbidity, colour and the presence of copper and mercury ions, the probe method is unaffected by these interferents. However, errors with the probe were evident in the analysis of calcareous soils due to carbon dioxide interference. The range of the method was 0.1–5 mg NO_2^- nitrogen per litre and the precision was excellent, with relative standard deviations being typically less than 1%.

As with the sulfur dioxide probe, it is necessary to carry out measurements in a closed system such as that shown in Fig. 7.4 because of the volatility of nitrogen oxides. For measurements below approximately 10^{-5} M nitrite, it is necessary to remove carbon dioxide from both samples and buffer to prevent interference; hence, measurement of nitrite in most natural water samples would require the use of a special monitoring system.

No applications of the hydrogen sulfide probe have yet been published.

REFERENCES

1. R. W. Stow, R. F. Baer and B. F. Randall, *Arch. Phys. Med. Rehabil.* **38**, 646 (1957).
2. W. Severinghaus and A. F. Bradley, *J. Appl. Physiol.*, **13**, 515 (1958).
3. W. Severinghaus, *Ann. N.Y. Acad. Sci.* **148**, 115 (1965).
4. J. Růžička and E. H. Hansen, *Anal. Chim. Acta* **69**, 129 (1974).
5. P. L. Bailey and M. Riley, *Analyst* **100**, 145 (1975).
6. J. W. Ross, J. H. Riseman and J. A. Krueger, *Pure Appl. Chem.* **36**, 473 (1973).
7. D. Midgley and K. Torrance, *Analyst* **97**, 626 (1972).
8. L. R. McKenzie and P. N. W. Young, *Analyst* **100**, 620 (1975).
9. R. F. Thomas and R. L. Booth, *Environ. Sci. Technol.* **7**, 523 (1973).
10. T. R. Gilbert and A. M. Clay, *Anal. Chem.* **45**, 1757 (1973).
11. M. J. Beckett and A. L. Wilson, *Water Res.* **8**, 333 (1974).
12. D. Midgley, *Analyst* **100**, 386 (1975).
13. U. Fiedler, E. H. Hansen and J. Růžička, *Anal. Chim. Acta* **74**, 423 (1975).
14. Orion Research Inc., *Instruction Manual, Sulphur Dioxide Electrode*, 3rd edn, 1974.
15. M. A. Tabatabai, *Soil Sci. Plant Anal.*, **5**, 569 (1974).
16. P. L. Bailey, unpublished work in the E.I.L. laboratory.
17. W. L. Banwart, M. A. Tabatabai and J. M. Bremner, *Soil Sci. Plant Anal.*, **3**, 449 (1972).
18. T. Almgren, *Anal. Chim. Acta* **73**, 420 (1974).
19. D. Midgley and K. Torrance, *Analyst* **98**, 217 (1973).
20. J. Mertens, P. Van den Winkel and D. L. Massart, *Bull. Soc. Chim. Belg.*, **83**, 19 (1974).
21. W. H. Evans and B. F. Partridge, *Analyst* **99**, 367 (1974).
22. H. F. Proelss and B. W. Wright, *Clin. Chem.* **19**, 1162 (1973).
23. I. Sekerka and J. F. Lechner, *Anal. Lett.* **7**, 463 (1974).
24. G. K. Buckee, *J. Inst. Brewing* **80**, 291 (1974).
25. J. Mertens, P. Van den Winkel and D. L. Massart, *Anal. Chem.* **47**, 522 (1975).
26. J. M. Bremner and M. A. Tabatabai, *Comm. Soil Sci. Plant Anal.* **3**, 159 (1972).
27. P. M. Todd, *J. Sci. Food Agric.* **24**, 488 (1973).
28. A. R. Deschrieder and R. Meaux, *Analusis* **2**, 442 (1973).
29. N. J. Park and J. C. B. Fenton, *J. Clin. Pathol.* **26**, 802 (1973).
30. G. T. B. Sanders and W. Thornton, *Clin. Chim. Acta* **46**, 465 (1973).
31. S. Van-Vlasselaer and F. Crucke, *Ann. Pharm. Fr.* **31**, 769 (1973).
32. M. L. Eagan and L. Dubois, *Anal. Chim. Acta* **70**, 157 (1974).
33. C. P. Sloan and G. P. Morie, *Anal. Chim. Acta* **69**, 243 (1974).
34. C. R. Caflisch and N. W. Carter, *Anal. Biochem.* **60**, 252 (1974).
35. J. A. Krueger, *Anal. Chem.* **46**, 1338 (1974).
36. P. L. Bailey, *J. Sci. Food Agric.* **26**, 558 (1975).
37. E. H. Hansen and J. Růžička, *Anal. Chim. Acta* **72**, 353 (1974).

CHAPTER 8

MISCELLANEOUS SENSORS

A number of ion-selective electrodes, or methods involving ion-selective electrodes, which do not comfortably fit into any of the other chapters are to be discussed in this chapter. Foremost among these are the enzyme electrodes and the concomitant methods of analysis involving both enzymes and ion-selective electrodes or gas-sensing probes.

ENZYME ELECTRODES AND ANALYSES WITH ENZYMES

Reactions involving enzymes may be represented in a simplified form by

$$\text{Substrate} \xrightarrow{\text{Enzyme}} \text{Products} \qquad (8.1)$$

Methods of analysis involving enzymes may be designed to measure either the substrate or the enzyme activity, and the sensor may be used to detect either one of the products or, less usually, the substrate. By combining the high selectivity of enzymatic reactions with the selectivity of ion-selective electrodes or gas-sensing probes, many excellent analytical methods may be developed.

Four essentially different types of procedure for performing these analyses may be distinguished.

1. *Single-stage bulk procedure*. The substrate and enzyme are mixed together and allowed either a standard time to react or sufficient time for the reaction to reach completion. The ion-selective electrode or gas-sensing probe immersed in the solution then indicates the activity of one of the reaction products or the decrease in substrate activity. The reaction may take place in a beaker or in an automated continuous flow system with a delay stage.

2. *Enzyme electrode procedure*. The enzyme (for substrate analysis) or the substrate (for enzyme analysis) is immobilized in a thin layer, often of gel, coated on to the surface of the sensor. The coated sensor is immersed in the

169

sample, which is stirred at a constant rate, and the cell potential is noted when it reaches equilibrium. A reaction occurs only within the coating and at the coating–sample interface.

3. *Double-stage bulk procedure.* This procedure is similar to procedure 1 except that the reaction and measurement stages are separated. In some cases the optimum sample pH for the greatest enzyme activity (i.e. the pH at which the reaction goes fastest) is outside the normal working pH range of the sensor. The reaction is therefore conducted at one fixed pH and the pH is then changed to a second value before the measurement is made.

4. *Multiple-stage bulk procedure.* If the product of an enzymatic reaction is not itself detectable by an electrode or probe, additional reactions, enzymatic or otherwise may be performed to generate a detectable species.

   Thus, for any particular enzyme-catalysed reaction, a large variety of analytical methods and procedures may be used, as the many papers on the urea/urease reaction demonstrate; these will not all be equally satisfactory and the best must be chosen for the analysis in hand. The first procedure discussed above was indeed the first to be developed, and was reported in its simplest form by Katz and Rechnitz[1] in 1963. They studied the analytical use of the urea/urease reaction and determined either urea or urease by means of an ammonium-selective glass electrode, which sensed the ammonium ions formed as product in the reaction

$$CO(NH_2)_2 + H_2O \xrightarrow{\text{Urease}} CO_3{}^{2-} + 2NH_4{}^+ \qquad (8.2)$$

The whole sample, of either substrate or enzyme, is mixed with a buffer solution and allowed to react to completion; hence, the procedure is relatively expensive because of the high consumption of reagents per analysis. This situation is aggravated when a standard ammonium-selective glass electrode is used as sensor, since the volume of sample plus reagent required is also quite large. A version of this procedure, for the analysis of urea, which requires only 0.1 ml sample and with the enzyme immobilized in the bottom of the sample chamber has been described by Guilbault and Tarp.[2] The reaction takes place in the chamber of an ammonia-sensing 'air-gap electrode' and, because of the relatively large surface area of the immobilized enzyme in contact with the small volume of sample, equilibrium is reached in 2–4 min. The major difficulty with any version of this first procedure is that the pH for the optimum enzyme activity frequently does not coincide with the optimum working range for the sensor. Thus, to take as an example the ubiquitous urea/urease reaction, the urease reaches maximum activity at about pH 7, whereas the most selective and satisfactory sensor for the reaction products, an ammonia-sensing probe, is best used in samples whose pH is greater than 12. In these cases use of procedure 3 is most satisfactory. Furthermore, since most reactions require a short incubation period, and thus the analyses cannot truly be performed in a single stage, the addition of a second chemical treatment stage is a negligible disadvantage.

The introduction of enzyme electrodes by Guilbault and his colleagues was an ingenious step to simplify analyses normally carried out by procedure 1. The true 'enzyme electrode', which is more widely used than the term 'enzyme substrate electrode' proposed by IUPAC,[3] may be defined as a sensor comprising an ion-selective electrode, or gas-sensing probe, the sensing surface of which is covered by a coating containing an enzyme, or substrate, which causes a reaction with the determinand substrate or enzyme to produce a species sensed by the sensor. An enzyme electrode is depicted in Fig. 8.1. In the more usual

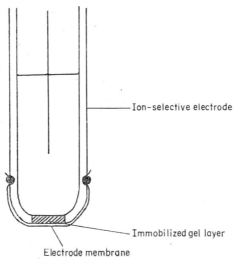

Fig. 8.1   An enzyme electrode

configuration the coating contains an enzyme either in suspension in a buffer, and held against the sensing surface by a dialysis membrane, or immobilized in a gel, typically a polyacrylamide gel, coated on to the sensing surface. The reaction occurs at the coating surface–sample interface and the sensor measures the equilibrium activity of the product, at the sensor surface, resulting from diffusion of the product through the coating.

There are several advantages and disadvantages of such enzyme electrodes. The advantages are clear. The electrode is convenient to use and, because only that part of the sample adjacent to the electrode surface reacts, only a small amount of enzyme is needed; furthermore, the enzyme is removed from the sample with the electrode after each analysis and may be used for many subsequent analyses. Sample pretreatment is minimized, as usually only addition of a suitable buffer is necessary.

The disadvantages of the electrodes are often less clearly stated in the literature. Their use is confined to reactions which can be followed by procedure 1, since both reaction and measurement take place simultaneously, and also to reactions which are relatively fast: these limitations are severe. The steady decrease of the enzyme activity produces drift in the standard potential of the electrode; such drift is usually rapid compared with the drift rate of most ion-selective electrodes or gas-sensing probes. In those electrodes where the enzyme is immobilized in a gel, the rate of diffusion of both products and substrate through the gel is slow and, hence, the response time of the electrode is slow and it shows memory effects; if however, an aqueous enzyme suspension is used as the coating, a much faster response is obtained but the steady dissolution of the enzyme, through the dialysis material into the sample, limits the lifetime to no more than a few days.[4] A further disadvantage of the enzyme electrodes is the sensitivity of the electrode response to the rate of stirring of the samples, which affects the diffusion equilibria at the coating surface: as pointed out by Nagy and Pungor,[5] the rapid attainment of equilibrium is dependent on the production of a thin diffusion layer of constant thickness, such as is produced by rapid stirring. Nagy and Pungor[5] suggest that the prime reason that enzyme electrodes are not generally used is because the difficulties of producing a reaction layer (coating) of sufficient activity and stability have not yet been overcome. An additional important factor is that the coating on the surface of the ion-selective electrode does not eliminate the interference with the electrode response caused by interferents in the sample. Thus, for example, the early urea electrodes (see, e.g., Ref. 6), made by coating ammonium-selective glass electrodes, suffered interference from sodium and potassium ions and thus could not be used for direct measurements of urea in clinical samples. Removal of the interferents by mixing with an ion exchanger[7] was possible but not easy.[2] The alternative electrode made by coating a monactin/nonactin silicone rubber based ammonium-selective electrode was also insufficiently selective.[8,9] The most selective system was that reported by Anfalt et al.,[4] who coated the membrane of an ammonia-sensing probe. However, at the pH necessary to produce adequate enzyme activity, only a small fraction of the ammonia species produced were present as the free gas and the sensitivity and the response time of the probe suffered; moreover, the system was extremely sensitive to pH changes in the sample, as this affected both the ratio of ammonia to ammonium ions in the coating and the activity of the enzyme. Thus, to summarize, even for the relatively simple urea/urease reaction it is difficult to construct a wholly satisfactory enzyme electrode, and it may often be better to use one of the other procedures.

The third and fourth procedures described above are the most versatile and will probably become the most widely used, especially in automated systems. An example of procedure 3 is the determination of urea described by Llenado and Rechnitz.[10] In their continuous-flow system the urea sample is mixed with a solution of urease in a phosphate buffer at pH 7.4 and then passes through an incubation coil held at 37 °C for 10 min; the reaction is then quenched by addition of 0.1 M sodium hydroxide solution and the alkaline sample, after debubbling, is passed through a flow-through cap fitted on to an ammonia-sensing membrane probe. The method was shown to be reproducible (standard deviations ranged from 0.3% at 4.7×10^{-2} M urea to 2.6% at 4.7×10^{-4} M urea), although the relative errors were quite large ($+14.7\%$ at the lowest concentration). In this method both reaction and measurement took place under optimum conditions; however, relatively large amounts of the enzyme are consumed and the accuracy is not very good, although it could almost certainly be substantially improved by thermostatting the probe and removing the debubbler. In an earlier paper Llenado and Rechnitz[11] describe in useful detail the parameters associated with the construction of a continuous-flow analyser employing enzymes.

The determination of total cholesterol in blood serum by use of an iodide-selective electrode according to the method of Papastathopoulos and Rechnitz[12] involves two enzyme reactions, and is thus an example of procedure 4. The reaction may be represented by the equations

$$\text{Cholesterol esters} + H_2O \xrightarrow{\substack{\text{Cholesterol} \\ \text{ester hydrolase}}} \text{free cholesterol} + \text{fatty acids} \quad (8.3)$$

$$\text{Free cholesterol} + O_2 \xrightarrow{\text{Cholesterol oxidase}} \text{cholest-4-en-3-one} + H_2O_2 \quad (8.4)$$

$$H_2O_2 + 2I^- + 2H^+ \xrightarrow{\text{Mo(VI)}} I_2 + H_2O \quad (8.5)$$

The reactions were carried out in the continuous-flow system shown in Fig. 8.2. The samples are mixed with a solution of the two enzymes in a phosphate buffer at pH 6.8 (the 'reagent' in the diagram) and segmented with a stream of oxygen; they are then passed through an incubation coil for 13 min at 37 °C. The reaction is quenched by addition of strong perchloric acid to give an acid concentration of approximately 0.5 M, and mixed with a solution containing iodide ions and,

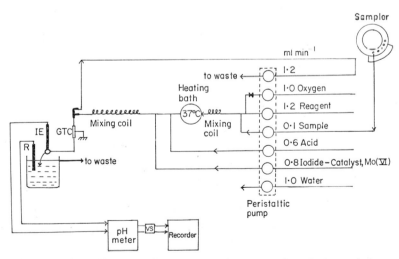

Fig. 8.2 Flow diagram of an automatic system for cholesterol determination in serum[12]

as catalyst, molybdenum(VI) (in the form of ammonium molybdate). The iodide electrode is used to measure the decrease in iodide concentration, which is in turn proportional to the total cholesterol concentration. Samples containing total cholesterol in the range 80–240 mg dl^{-1} were analysed at the rate of 20 per hour with relative errors of 3% or less and an average relative standard deviation of 2.6%. Very satisfactory agreement was obtained between this method and two spectrophotometric methods.

A summary of some of the more useful methods of analysis involving enzymes and ion-selective electrodes and gas-sensing probes is given in Tables 8.1 and 8.2: more comprehensive but less selective lists have recently been presented by Nagy and Pungor[5] and Moody and Thomas.[13] A review of methods involving enzymes and voltammetric sensors is also included by Nagy and Pungor.[5]

The choice of reference electrode for use in conjunction with ion-selective electrodes needs special consideration in these applications, since many enzyme reactions are severely inhibited by trace amounts of silver or mercury ions. Hence, it is advisable in manual methods to use a double junction reference electrode with a normal bridge solution such as 3.8 M potassium chloride solution. In continuous-flow systems the sample does not usually pass the reference electrode until it has passed the sensing electrode and until the enzyme reaction has been quenched, so no special precautions are necessary.

CHALCOGENIDE GLASS ELECTRODES

Ion-selective electrodes sensing ferric and cupric ions have been prepared from chalcogenide glasses. These curious electrodes were first reported by Baker and Trachtenberg.[22] They suffer too many disadvantages for analytical use at the moment but are nevertheless electrochemically interesting.

The ferric ion-selective electrode[23] is prepared from the glass $Fe_nSe_{60}Ge_{28}Sb_{12}$. The electrode calibration graph shows a slope of 60 mV decade^{-1} in the range 10^{-5}–10^{-1} M: this slope, which indicates a one-electron transfer process, and other data suggest that the electrode responds by means of a redox mechanism on the electrode surface. The electrode does not respond to ferrous ions but suffers severe interference from oxidants stronger than ferric ions and also silver ions. Unfortunately, the electrode requires frequent reactivation by etching the surface and oxidizing it in a 10^{-3} M ferric ion solution: thereafter deactivation sets in, at a rate proportional to the average ferric ion concentration to which the electrode is exposed, and is very fast at 10^{-1} M. The response time of the electrode was excessively long by the standard of other electrodes and probes; the electrode took 10–20 min to reach a steady state after a change to 10^{-4} M ferric ions.

The cupric ion-selective electrode[24] is subject to the same drawbacks as the ferric ion-selective electrode except that it has a faster response: it is made from a $Cu–As_2S_3$ glass in which the active species is reported to be the material known as sinnerite ($Cu_6As_4S_9$). The electrode calibration graph has a slope of

TABLE 8.1
Enzymatic methods of analysis for substrates

Determinand	Enzyme	Sensor	Concn range	Reaction pH	Measurement pH	Comments	Ref.
Urea	urease	ammonia-sensing membrane probe	10^{-4}–10^{-1} M	7.4	12	Continuous flow method.	10
Urea	urease	ammonia 'air-gap' electrode	10^{-4}–10^{-2} M	7.0	12	Manual method.	14
Creatinine	creatininase	ammonia-sensing membrane probe	1–100 mg %	8.5	12	Procedure 3, semi-automated.	15
Amygdalin	β-glucosidase	cyanide electrode	10^{-4}–10^{-1} M	7	7	Enzyme electrode method.	16
Cholesterol	(i) cholesterol ester hydrolase (ii) cholesterol oxidase catalyst: Mo^{VI}	iodide electrode	80–420 mg dl^{-1}	6.8	<1	Treble-stage procedure including two enzyme reactions.	12
Penicillin	Penicillinase	pH electrode	3.5–1100 mg l^{-1}	—	—	Enzyme electrode method (Rel. std deviation 3 % at 100 mg l^{-1}).	19
L-Phenylalanine	L-amino acid oxidase	ammonium electrode (monactin/nonactin)	5×10^{-5}–10^{-2} M	7	7	Enzyme electrode method. Precision, about 1.7%; accuracy, 2%.	20
Glucose	(i) glucose oxidase (ii) peroxidase	iodide electrode	10^{-4}–10^{-3} M	5 ± 0.5	5 ± 0.5	Flow system with enzyme electrode. Two consecutive reactions. Slow response.	21

TABLE 8.2
Methods of analysis for enzymes

Determinand	Substrate	Sensor	Concn range	Reaction pH	Measurement pH	Comments	Ref.
Acetylcholinesterase	acetylcholine	choline ester electrode	0.2–2.5 i.u.	7.0	7.0	Hydrolysis of the substrate is monitored.	17
Nitrate reductase	10^{-3} M nitrate + 8×10^{-2} M formate	nitrate electrode	0.007–0.7 i.u.	7.2	7.2	Flow system used.	18
β-Glucosidase	amygdalin	cyanide electrode	0.025–1.3 units	6.2	12	Continuous flow method; 20 samples per hour.	11
Rhodanase	thiosulfate/cyanide	cyanide electrode	3–30 units	7.9	12	Continuous flow method; 20 samples per hour.	11
Glucose oxidase	glucose	iodide electrode	0.02–1.9 units	5.7	12	70 samples per hour.	11

approximately $29\,mV$ decade^{-1} between 10^{-2} and $10^{-6}\,M$, indicating a response mechanism involving a two-electron exchange. The electrode did not respond satisfactorily, and also partially dissolved, in halide solutions. The cupric ion-selective electrode described in Chapter 5 is more suitable for analytical work, although this also does not respond satisfactorily in halide solutions.

SURFACTANT ELECTRODES

Liquid ion-exchange electrodes responding to cationic[25] and anionic[26] surfactants were first reported by Gavach and co-workers. The ion exchangers for the cationic surfactants[25] were prepared by dissolving the tetraphenylborate or picrate salt of the ion to be measured in nitrobenzene. Thus, a satisfactory electrode for measuring dodecyltrimethylammonium ions down to $10^{-6}\,M$ may be prepared from dodecyltrimethylammonium tetraphenylborate; the cation may be measured either directly or by means of a potentiometric titration using sodium tetraphenylborate as titrant. The anionic detergents[26] dodecylsulfate, tetrapropylenebenzenesulfonate and dioctylsulfosuccinate could be measured down to 10^{-5}, 5×10^{-6} and $10^{-6}\,M$ respectively and up to close to the critical micelle concentration for each species. Mixtures of the three could also be determined by potentiometric titration using the dodecylsulfate electrode to detect the end-points.

Birch and Clarke[27] later showed that a more satisfactory electrode may be prepared from hexadecylpyridinium dodecylsulfate dissolved in 0.17 M hexachlorobenzene + 0.017 M bromoacetanilide in o-dichlorobenzene: use of this ion-exchanger allowed accurate determination of the critical micelle concentration of the dodecylsulfate anion.[28,29] The electrode measured dodecylsulfate ion concentrations down to $10^{-6}\,M$ (limit of Nernstian response, $3 \times 10^{-5}\,M$) in the pH range 4–10. Anions with a shorter chain length than dodecylsulfate did not interfere, but ions such as tetradecylsulfate and hexadecylsulfate did interfere: the selectivity coefficient could be related to the ratio of the partition coefficients of the determinand and interferent between the aqueous sample and organic ion-exchanger phases. The electrode could also be used to measure cationic surfactants such as tetradecyltrimethylammonium ions and their critical micelle concentrations.

The dodecylsulfate electrode was later used by Birch et al.[29] to study the binding of dodecylsulfate ions to non-ionic polymers such as Dextran (T-70), polyvinyl alcohol (PVA) and polyvinylpyrrolidone (PVP). The effect of protein (bovine serum) albumin, on the dodecylsulfate activity and electrode response, was also investigated. In all cases the electrode gave reliable and rapid data on the determinand activity.

A version of this system with the ion exchanger, hexadecyltrimethylammonium dodecylsulfate, incorporated in silicone rubber, has been reported by Fogg, Pathan and Burns:[30] the preparative details for the membrane are given in full.

The electrode responded to cationic surfactants (hexadecyltrimethylammonium ions and hexadecylpyridinium ions) but, curiously, not to anionic detergents. These electrodes show strong memory effects and appear to be less satisfactory in general use than the electrodes previously described.

ACETYLCHOLINE ELECTRODE

An electrode to measure acetylcholine has been described by Baum;[31] the original form of the electrode with a liquid membrane could be used to measure acetylcholine in the concentration range 10^{-1}–10^{-7} M (limit of Nernstian response, 10^{-5} M). The electrode showed high selectivity over potassium, sodium and ammonium ions, and also good selectivity over choline ($k_{ACh.Ch} = 0.066$). The electrode was used[32] to study the reaction

$$\text{Acetylcholine} \xrightarrow{\text{Acetylcholinesterase}} \text{Choline} \qquad (8.6)$$

The reaction proceeded after a short induction period; the rate of change of electrode potential was proportional, during the first 40% of the hydrolysis, to the activity of the enzyme acetylcholinesterase present. Thus, the electrode could be used for the indirect determination of the enzyme in the activity range 0.2–8 units. The extent of the hydrolysis which could be followed satisfactorily was limited by the eventual interference caused by the choline produced. The results agreed well with those from a spectrophotometric method. The response time of the electrode is quoted as 15–30 s.

Baum et al.[17] later reported a version of the electrode with the ion exchanger, acetylcholine tetra-p-chlorophenylborate, dissolved in a phthalate ester immobilized in PVC. This electrode showed much inferior selectivity ($k_{ACh.Ch} \simeq 0.2$): when used for the assay of acetylcholinesterase, the measurements did, however, show less scatter than those with the liquid membrane electrode. The electrode was also shown to be very sensitive to other choline esters such as acetyl-β-methylcholine and, particularly, butyrylcholine. The selective measurement of this latter choline ester is apparently possible with the electrode, since the selectivity coefficients are in its favour; with respect to choline, the quoted values of the coefficient are $K_{BCh.Ch} \simeq 0.02$ (for PVC membranes) or 0.0025 (for liquid membranes).

CHIRALITY-SENSING ELECTRODE

Thoma et al.[33] have described an interesting electrode which is able to distinguish between the enantiomers of α-phenylethylammonium ions. This electrode, of which the active component is a chiral, electrically neutral carrier compound, gave potentials differing by approximately 2.2 mV when immersed in solutions of each of the enantiomers in turn at the same concentrations in the range 10^{-1}–10^{-3} M. The samples were buffered to pH 4.4 by means of an acetate buffer. Dialysis experiments and potentiometric measurements showed that the selectivity coefficient of the electrode for one enantiomer over the other was 1.08.

ION-SENSITIVE FIELD EFFECT TRANSISTORS

Interesting new additions to the range of types of sensor which may be used to measure the activities of ions in solution are the ion-sensitive field effect transistors (ISFETs). The first of these devices was described by Bergveld[34] in 1972; more recently a review[35] has been published of the few papers which relate to this field.

Moss, Janata and Johnson[36] have studied the construction and performance of a potassium ISFET. In this device the metal gate of a conventional metal oxide semiconductor field effect transistor (MOSFET) is replaced by a potassium-selective membrane consisting of PVC incorporating valinomycin (see Fig. 8.3).

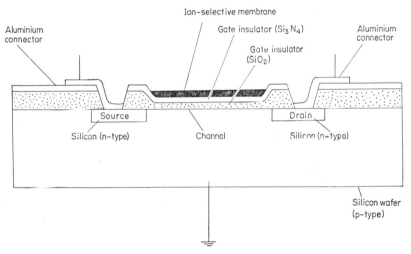

Fig. 8.3   An ion-sensitive field effect transistor[36]

The potential developed across this membrane regulates the potential on the surface of the gate insulator and, hence, the current flowing through the channel between the drain and the source. The current flowing is proportional to the logarithm of the activity of the determinand, in this case potassium ions. The device is sealed into the end of a tube with just the membrane exposed to the solution, and is used in a conventional manner in conjunction with a reference electrode: earlier versions were reputed to work without reference electrodes, but Moss et al.[36] suggest that leakage of solution through the insulation round the ISFET effectively produced a reference system. The sensor in its present form does not have such a long linear calibration range as a normal potassium-selective electrode based on valinomycin, its useful life is restricted to one week and, moreover, it requires a lengthy activation period. Its advantages, on the other hand, are that electrical noise due to pick up is eliminated, as impedance conversion and signal amplification occur in the sensor tip. These advantages do not sufficiently offset the disadvantages of the sensor in its present state. However, it is predicted that improvements will soon be forthcoming, and these are awaited with interest.

REFERENCES

1. S. A. Katz and G. A. Rechnitz, *Z. Anal. Chem.* **196**, 248 (1963).
2. G. G. Guilbault and M. Tarp, *Anal. Chim. Acta* **73**, 355 (1974).
3. *Recommendations for Nomenclature of Ion-Selective Electrodes*, Appendices on Provisional Nomenclature, Symbols, Units and Standards—Number 43, IUPAC Secretariat, Oxford (January 1975).
4. T. Anfalt, A. Graneli and D. Jagner, *Anal. Lett.* **6**, 969 (1973).
5. G. Nagy and E. Pungor, *Hung. Sci. Inst.* No. 32, 1 (1975).
6. G. G. Guilbault and J. G. Montalvo, *Anal. Lett.* **2**, 283 (1969).
7. G. G. Guilbault and E. Hrabankova, *Anal. Chim. Acta* **52**, 287 (1970).
8. G. G. Guilbault and G. Nagy, *Anal. Chem.* **45**, 417 (1973).
9. G. G. Guilbault, G. Nagy and S. S. Kuan, *Anal. Chim. Acta* **67**, 195 (1973).
10. R. A. Llenado and G. A. Rechnitz, *Anal. Chem.* **46**, 1109 (1974).
11. R. A. Llenado and G. A. Rechnitz, *Anal. Chem.* **45**, 826 (1973).
12. D. S. Papastathopoulos and G. A. Rechnitz, *Anal. Chem.* **47**, 1792 (1975).
13. G. J. Moody and J. D. R. Thomas, *Analyst* **100**, 609 (1975).
14. E. H. Hansen and J. Růžička, *Anal. Chim. Acta* **72**, 353 (1974).
15. H. Thompson and G. A. Rechnitz, *Anal. Chem.* **46**, 246 (1974).
16. M. Mascini and A. Liberti, *Anal. Chim. Acta* **68**, 177 (1974).
17. G. Baum, M. Lynn and F. B. Ward, *Anal. Chim. Acta* **65**, 385 (1973).
18. W. R. Hussein and G. G. Guilbault, *Anal. Chim. Acta* **76**, 183 (1975).
19. L. F. Cullen, J. F. Rusling, A. Schleifer and G. J. Papariello, *Anal. Chem.* **46**, 1955 (1974).
20. G. G. Guilbault and G. Nagy, *Anal. Lett.* **6**, 301 (1973).
21. G. Nagy, L. H. von Storp and G. G. Guilbault, *Anal. Chim. Acta* **66**, 443 (1973).
22. C. T. Baker and I. Trachtenberg, *J. Electrochem. Soc.* **118**, 571 (1971).
23. R. Jasinski and I. Trachtenberg, *J. Electrochem. Soc.* **120**, 1169 (1973).
24. R. Jasinski, I. Trachtenberg and G. Rice, *J. Electrochem. Soc.* **121**, 363 (1974).
25. C. Gavach and P. Seta, *Anal. Chim. Acta* **50**, 407 (1970).
26. C. Gavach and C. Bertrand, *Anal. Chim. Acta* **55**, 385 (1971).
27. B. J. Birch and D. E. Clarke, *Anal. Chim. Acta* **67**, 387 (1973).
28. B. J. Birch and D. E. Clarke, *Anal. Chim. Acta* **61**, 159 (1972).
29. B. J. Birch, D. E. Clarke, R. S. Lee and J. Oakes, *Anal. Chim. Acta* **70**, 417 (1974).
30. A. G. Fogg, A. S. Pathan and D. T. Burns, *Anal. Chim. Acta* **69**, 238 (1974).
31. G. Baum, *Anal. Lett.* **3**, 105 (1970).
32. G. Baum, *Anal. Biochem.* **39**, 65 (1971).
33. A. P. Thoma, Z. Cimerman, U. Fiedler, D. Bedeković, M. Güggi, P. Jordan, K. May, E. Pretsch, V. Prelog and W. Simon, *Chimia* **29**, 344 (1975).
34. P. Bergveld, *IEEE Trans.* **BME-19**, 342 (1972).
35. J. N. Zemel, *Anal. Chem.* **47** (2), 255A (1975).
36. S. D. Moss, J. Janata and C. C. Johnson, *Anal. Chem.* **47**, 2238 (1975).

CHAPTER 9

PRACTICAL PROCEDURES

In this chapter some of the practical procedures referred to in previous chapters are to be described in more detail. The practical points to be considered in the development of a method using an ion-selective electrode or gas-sensing probe for either laboratory or plant applications will be considered in turn and the methods compared with those using other measurement techniques such as spectrophotometry.

ESSENTIAL EQUIPMENT

One of the attractive features of ion-selective electrodes or gas-sensing probes is that they do not have to be used in conjunction with a clutter of expensive equipment and accessories.

The prime requirement is a pH/millivoltmeter which can be read to ± 0.2 mV or better. The high-impedance socket into which the electrode or probe is plugged should have an impedance of greater than 10^{12} Ω; this is particularly important when gas-sensing probes are used, as these usually have a higher impedance than the other sensors. This impedance requirement is met by the meters of all reputable manufacturers; nearly all the meters use comparable field effect transistor input stages.

The introduction of the new generations of ion-selective electrodes and gas-sensing probes has led to the incorporation of two important features additional to those of the traditional design specification of pH/millivoltmeters: these features, double high-impedance inputs and direct-reading concentration scales, are slowly but surely becoming more common.

A few meters are now available with double high-impedance inputs, so that two high impedance electrodes may be used in the cell. This is particularly useful for measurements in cells without a liquid junction; for example, a cell could be constructed to measure fluoride ions with a pH electrode (monitoring the constant pH fixed by the T.I.S.A.B. reagent) as reference electrode.

The second feature, which is now fairly common, is meter scales marked out antilogarithmically for direct concentration measurements: an example is shown in Fig. 9.1. These direct-reading concentration scales enable the meter to be

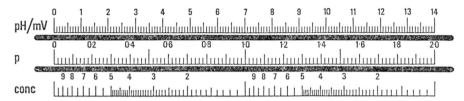

Fig. 9.1   The scale of an expanded range pH/millivolt meter, including a direct reading concentration scale (E.I.L. Model 7050 pH meter)

calibrated in concentration units so that measurements may be made without recourse to the usual calibration graph of potential against log concentration (or activity). Such concentration scales must be used with caution, since the actual potential values give valuable information to the analyst on the perform- ance of the cell; a sudden or drifting change in the standard potential of the cell, which might easily pass unobserved when calibrating directly in concentration units, may presage failure of a cell component. The direct-reading scales presume a linear potential against log concentration calibration plot, and therefore should not be used when the concentration, of either samples or any of the calibration standards, is likely to fall below the limit of Nernstian response of the electrode or probe. If one of the standards is below this limit, a warning will be given by the value of the calibration slope set on the meter (most meters allow a reading to be made of the % theoretical slope or a related parameter), since, in order to calibrate the concentration scale using the standards, this control will have to have been set at less than the theoretical value. In such cases the measurements must be made by means of a calibration graph. However, for the vast majority of satisfactory applications the concentration range of the samples and standards lies above the limit of Nernstian response, and it is very con- venient and economical of time to use these direct-reading scales. The necessity of plotting graphs is eliminated and the whole procedure is made both faster and simpler.

Another feature, offered by Orion Research Inc. on some of their portable pH meters, is scales for directly reading the results of analyses performed by standard addition or standard subtraction methods. These scales can only be used if the procedure laid down by Orion is exactly followed and if the procedure is suitable for the analysis is question. However, this facility is very convenient for the determination of a species which is largely complexed in the samples and which cannot be easily decomplexed by pretreatment for direct measurement, and when high accuracy or precision are not of paramount importance. For accurate and precise determinations it is preferable to use a conventional pH/millivoltmeter and measure the actual cell potential.

Other equipment, for use with ion-selective electrodes or gas-sensing probes, ensures that the sample is presented to the sensor in a uniform and satisfactory manner. Since the potentials of many electrodes, particularly those with liquid membranes, and the liquid junction potentials of reference electrodes, depend on the velocity of the sample past the electrode, all samples should be stirred: for the most precise results the stirring speed and the disposition of electrodes in the sample container should be kept constant. These conditions are usually assured if the electrodes are used in a continuous-flow system. Samples contained in beakers may conveniently be stirred by means of a magnetic stirrer and stirrer bars. Many poor-quality stirrers generate heat and stir faster as the stirrer warms up; neither fault can be tolerated. The former fault is the more common and many, if not most, commercial stirrers become sufficiently warm to raise the temperature of small samples by several degrees Celsius over five minutes: this effect may be partly, and sometimes sufficiently, suppressed by placing an asbestos square on top of the stirrer to insulate it from the samples. Care should be taken to ensure that the coating on the stirrer bars is resistant to the samples and reagents and that the bars are washed between measurements.

In general, no other equipment is required for direct measurements, but for the most accurate and precise work it is necessary to thermostat both samples and standards before measurement.

CALIBRATION

After the necessary equipment and electrodes for the determination to be performed have been accumulated, the next stage is usually to calibrate the cell. Calibration is essential when direct methods of analysis (described in the next section) are used; it is also strongly advised for standard addition procedures, since the actual slope of the calibration graph is required in the calculation and this may differ significantly from the theoretical slope. The only major area of application in which calibration is not necessary is when the sensor is used to detect the end-points of potentiometric titrations.

In Chapter 3 the theoretical considerations associated with calibration procedures were outlined; in particular, the choice between calibrating with activity or concentration standards was discussed. In each case calibration may be done either by a manual method or by an automatic or semi-automatic method.

The manual method of calibration is extremely simple. First a series of solutions is prepared spanning the anticipated range of activity or concentration of the determinand in the samples. If the calibration graph is expected to be linear with close to the theoretical Nernstian slope in the measurement range and the range is broad, it is usually sufficient to prepare one standard per decade of activity or concentration. If, however, the measurement range is very narrow (<0.5 decade) or the calibration graph is curved, the standards should bracket the range as closely as possible, and as many intermediate standards should be used as are commensurate with the accuracy or precision required and the degree of curvature of the graph. So, after selection of the calibration standards, an aliquot of each is pretreated with whatever reagent, if any, is appropriate to the method and brought to the same temperature as the samples. The ion-selective electrode and reference electrode, or gas-sensing probe, are then immersed in the pretreated and continuously stirred standards in ascending order of activity (or concentration). The potential of the cell is measured in each solution and the resultant potentials are plotted against the logarithm of the determinand activity (or concentration) to produce a calibration graph. The reproducibility of the cell may then be checked by reversing the procedure and measuring the cell potentials in decreasing order of solution activity (or concentration). When a pH meter with a direct-reading concentration scale (see previous section) is used, only two standards at the extremes of the range of interest are necessary to calibrate the meter but a third is useful to check the linearity of the cell response. The standards used in these procedures may be either activity or concentration standards prepared by serial dilution or solutions buffered with respect to the determinand.

A convenient and rapid method for preparing a calibration graph with concentration standards, by a reversal of the serial dilution technique, is known as the 'litre beaker method'.[1] The method is very suitable for calibrations at the lower end of the response range of an electrode, where errors of a few per cent may be tolerated. One litre of sample background is mixed with any sample pretreatment reagent required and then transferred to a one litre beaker (a larger beaker may be necessary if the ratio of the volume of pretreatment reagent to sample is large); a plastics beaker may be appropriate for calibrations at very low determinand concentrations to minimize errors due to adsorption. The solution is stirred at a fixed rate and the ion-selective electrode and reference electrode are immersed in the solution. Since the reference electrode remains immersed in this solution throughout the whole calibration, it is particularly important to use a bridge solution which does not contaminate the solution. Small increments of a strong standard solution of the determinand are then pipetted into the solution to produce the required concentrations and the cell potentials are recorded after each addition. The dilution produced is normally ignored up to a total volume of added standard of 10 ml (1% error) as the method is essentially rapid and approximate, rather than highly accurate, and the electrode response is in any case less reproducible at low concentrations. If a fluoride electrode is to be calibrated for the analysis of potable waters, for example, the fluoride electrode and reference electrode are immersed in 1 litre of distilled water to which is added 100 ml concentrated T.I.S.A.B. solution; 0.1 ml of a standard 1000 mg l^{-1} standard fluoride solution is then added from a micropipette or 1 ml graduated pipette to produce a 0.1 mg l^{-1} treated fluoride standard. The cell potential in this standard is allowed to stabilize and is noted; successively larger increments of the standard are then added until the final total volume of standard added is 10 ml. The electrodes are thus standard-ized for the analysis of samples containing fluoride in the concentration range 0.1–10 mg F^- per litre.

Instead of adding the standard increments of determinand to the solution by pipette, the determinand may be introduced more accurately and precisely, and without dilution errors, by coulometric generation from electrodes immersed in the solution. This method, proposed by Bailey and Pungor,[2] was successfully applied to the calibration of iodide and silver ion-selective electrodes in the concentration range 10^{-4}–10^{-7} M, but may also be used for other determinands such as fluoride, bromide and chloride which can be coulometrically generated with 100% efficiency. Unfortunately, the number of those determinands which can be so generated is small. The method is primarily applicable, as is the 'litre beaker method', for low concentrations of determinand. It may readily be adapted for the automatic calibration of electrodes in automatic analysers.

A further calibration method which may be automated is that described by Horvai, Tóth and Pungor.[3] In this method the electrodes are calibrated in the reverse direction to the usual, starting with the strongest solution and finishing with the weakest solution. The electrodes are first immersed in a vessel com-pletely filled with a known volume of standard solution, the concentration of which is the top limit of the desired calibration range: the vessel is fitted with an overflow to keep the solution volume constant during the calibration. Back-ground electrolyte not containing the determinand is pumped into the vessel, either continuously or periodically, and thus gradually dilutes the standard; the solution is kept homogeneous by vigorous stirring but the volume remains constant as excess solution is discharged via the overflow. The concentration of determinand in the vessel at any instant may be calculated from the total pump-ing time, the flow rate, the volume of the vessel and the initial determinand concentration; thus, a calibration curve may be plotted from the calculated concentrations and corresponding measured potentials. The method may also be used to determine selectivity coefficients by pumping a solution containing a high concentration of interferent plus the standard concentration of deter-minand. This will produce a gradual increase in the interferent concentration and enables a conventional graph to be constructed for calculation of the selectivity coefficient (see Fig. 3.3).

MEASUREMENT TECHNIQUES

Many measurement techniques involving ion-selective electrodes or gas-sensing probes have been described; which of them is chosen for a particular application will depend upon such parameters as the composition of the samples, the required precision and accuracy and the time available to perform each analysis. The most important and widely used techniques are as follows: (i) direct method, (ii) standard addition/subtraction methods, (iii) Gran's plot, (iv) potentiometric titrimetry. These are listed in increasing order of accuracy and precision and also in increasing order of time required per analysis. Generally, the usefulness of each technique is inversely proportional to its degree of complication and the length of text required to explain it. However, not all techniques are applicable to all analyses: for example, a potable water sample containing 10^{-5} M fluoride could not be analysed by potentiometric titrimetry, because there is no suitable titrant, but could be analysed by any of the other three techniques. Each of the above four techniques will be considered in turn.

Direct method

The direct method is, as the name implies, the most straightforward technique and is the technique of choice whenever possible. The gas-sensing probe or ion-selective electrode and reference electrode are calibrated as described in the previous section; this results in the production of a calibration curve or in the calibration of a pH meter scale. The sample to be analysed is pretreated as required and the electrodes or probe are immersed in it. The equilibrium cell potential is then measured and related to the determinand activity or concentration by means of the calibration graph: alternatively, when a pH meter with a direct-reading concentration scale is used, the determinand activity (or concentration) may be read directly from the meter scale.

This technique is usually extremely rapid enabling measurements to be completed in just two or three minutes; in continuous-flow systems up to about 100 samples per hour may be analysed. The technique is also the most familiar and simple to workers not conversant with the newer generations of potentiometric sensors, since it is fundamentally identical with the commonest technique for pH measurement. The typical accuracy of measurement is approximately $\pm 2\%$ (± 0.5 mV for monovalent determinands) for samples in beakers at room temperature at concentrations above the limit of Nernstian response of the sensor. In less favourable experimental conditions, such as when interferents are present or the samples are warm, the accuracy and precision may be considerably worse. Conversely, if great care is taken over temperature control, constant stirring rates, sample pretreatment, etc., the accuracy may be improved to $\pm 1\%$: for the most accurate work the cell should be frequently checked with a single standard to ensure that the calibration is still valid. A large majority of the ion-selective electrode applications and virtually all the gas-sensing probe applications which have been described involve this technique. It is a major part of the attraction of the electrodes and probes that they produce results rapidly and with accuracy sufficient for most purposes: as soon as the technique is elaborated to give greater accuracy the speed is lost and the technique may often lose its advantage over alternative methods.

The direct method is suitable for the analysis of all samples in which the determinand is present in the 'free' (uncomplexed) state or in which the determinand may be 'unbound' (decomplexed) by appropriate pretreatment.

The electrode may also be used for the analysis of samples in which only a proportion of the determinand is 'free', provided that care is taken to ensure that that proportion is constant in all samples and standards. An example of such an application is the direct determination of calcium in waters described by Hulanicki and Trojanowicz[4] and previously referred to on p. 137. One of the aims of the pretreatment reagent in this case is to mask magnesium ions, but this cannot be achieved without partial complexation of the calcium ions; an excess of complexant, iminodiacetic acid, is therefore added to the reagent to ensure that the proportion of the calcium ions which remain 'free' is constant. Provided that the standards are similarly treated and that the concentration remaining 'free' after complexation is still above the limit of Nernstian response of the calcium electrode, the method provides a rapid and accurate method for calcium determination. Inevitably methods such as this, involving partial complexation of the determinand, are less sensitive than corresponding methods in which all the determinand is decomplexed.

Standard addition/subtraction methods

In the standard addition method (alternatively called the 'known addition method') the ion-selective electrode and reference electrode are immersed in the sample and the equilibrium cell potential is recorded; a known volume of a standard solution of the determinand is then added and the new equilibrium cell potential, and, hence, the change in potential, measured. The initial concentration of determinand in the sample may then be calculated from the equation

$$[X]_A = \frac{[X]_B}{10^{\Delta E/S}[1 + (V_A/V_B)] - (V_A/V_B)} \tag{9.1}$$

where $[X]_A$ and $[X]_B$ are the concentrations of the determinand X in the sample and standard, respectively; V_A and V_B are the initial volume of sample and the added volume of standard; ΔE is the change in cell potential; and S is the slope of the electrode calibration plot. This equation is simply derived from the equations for the electrode potentials in the sample before and after addition of the aliquot of standard solution.[5] The standard or known subtraction method[6] is different only in that the standard solution added to the sample is not a determinand solution but a solution of a species which reacts quantitatively with the determinand. Thus, a decrease in determinand concentration is produced with corresponding change in cell potential; this potential change may be used to calculate the initial determinand concentration in the sample by means of Eq. (9.1) adapted by addition of minus signs in front of $[X]_B$ and ΔE. This assumes a 1:1 stoichiometry of the reaction between the determinand and the added species: the equation becomes more complex if there is a different stoichiometry.

Equation (9.1) is derived on the assumption that the activity coefficient of the determinand is identical before and after the addition/subtraction and that the degree of complexation of the determinand also remains constant throughout.

If, as is sometimes the case, the dilution caused by the addition of the standard solution is sufficiently small to be negligible, with regard to the required accuracy of the analysis and the relative volumes of the sample and the aliquot of standard, Eq. (9.1) may be reduced to the following forms:

$$\text{Standard addition} \qquad [X]_A = [X]_B \frac{V_B}{V_A} \frac{1}{10^{\Delta E/S} - 1} \tag{9.2}$$

$$\begin{array}{c}\text{Standard subtraction}\\ \text{(1:1 stoichiometry)}\end{array} \quad [X]_A = [X]_B \frac{V_B}{V_A} \frac{1}{1 - 10^{-\Delta E/S}} \tag{9.3}$$

In standard subtraction methods to determine divalent ions, such as sulfide ions, by addition of a standard solution of monovalent ions, such as silver ions, the stoichiometry of 1:2 alters Eq. (9.3) to

$$[X]_A = [X]_B \frac{V_B}{2V_A} \frac{1}{1 - 10^{-\Delta E/S}} \tag{9.4}$$

Several assumptions are made in the derivation of these equations, as already stated: it is worth while listing these, and further assumptions which may be made to simplify the procedure or the calculation,[7,8] and seeing how the conditions of measurements should be fixed to make them justifiable.

(1) The change in sample volume caused by the addition of the aliquot is negligible (Eqs 9.2–9.4 only).
(2) The change in the liquid junction potential of the reference electrode is negligible.
(3) The activity coefficient of the determinand remains constant.
(4) The degree of complexation of the determinand remains constant.
(5) No interferents are present in the sample.
(6) The ion-selective electrode has a theoretical Nernstian response slope.
(7) In known subtraction methods the final concentration of determinand is above the limit of Nernstian response of the electrode.

Of these assumptions the first is seldom true since, typically, the volume V_B of the added aliquot of standard solution is at least 1–2%, and often 10%, of the initial sample volume V_A. The second and third assumptions are usually sufficiently true if an ionic strength adjustment buffer is added to the samples before measurement; as an added precaution the standard solution may also be prepared in this buffer, so that the composition of the sample background remains absolutely constant throughout the experiment. The validity of the fourth and fifth assumptions depends on the sample composition. Where necessary, interferents should be removed by sample pretreatment. If the determinand is completely uncomplexed ('free'), then the fourth assumption holds. However, if the determinand is partly complexed, whether by components of the sample or by the pretreatment reagent, there must be a large excess (50–100-fold ideally) of complexant present to ensure that the proportion of complexed species is held constant before and after the standard additions; in some cases it is necessary to add excess complexant to ensure this condition. The degree of complexation is often very pH-dependent; hence, buffering is to be recommended (if not already used to adjust the pH to the optimum working range of the electrode). Many pH buffers, such as acetate and phosphate, will complex many determinands and thus perform a dual role. However, the degree of complexation should not be so great that the free determinand activity is below the limit of Nernstian response of the electrode. The sixth assumption, that the electrode has a Nernstian response, is perhaps the most dangerous and it is always advisable to measure the actual slope for substitution into the equation; otherwise, inaccuracies of several per cent may frequently be introduced. Unfortunately, the determination of the slope is tantamount to plotting a calibration graph and much of the advantage of using the standard addition/ subtraction method is thereby lost.

The major source of error in Eq. (9.1) lies in the measurement of ΔE,[8] since the other terms V_A, V_B, $[X]_B$ and S may all be determined relatively accurately and precisely (better than 1% in all cases). The potentials of the cell, before and after the addition, from which ΔE is calculated will usually each be measured to ± 0.1 or ± 0.2 mV; so in order to minimize the relative error, the amount of the standard added ($[X]_B \times V_B$) should be sufficient to produce a value of ΔE of no less than 5 mV, and preferably between 10 mV and about 30 mV. Because of considerations relating to assumptions 3 and 4 it is preferable to maximize $[X]_B$ and minimize V_B (without making it so small that the volume is imprecisely dispensed) for a particular value of $[X]_B \times V_B$.

After developing a method for the analysis of a particular type of sample, taking all the previously mentioned factors into account and fixing V_A, V_B, $[X]_B$ and S, the calculation of the results may be simplified by plotting a graph or constructing a table of $[X]_A$ against ΔE. This graph or table may then be used to read off the required concentration of determinand after measurement of ΔE. Provided the method is strictly adhered to and S remains constant, the graph or table may be used indefinitely, despite changes in the reference electrode potential or in the standard potential of the ion-selective electrode, which are the usual causes of error in direct methods. The use of standard addition meter scales, such as on one Orion meter, considerably accelerates the analyses, as previously mentioned.

The standard addition/subtraction method is useful for two types of analysis— the rapid and approximate analysis of occasional samples and, more importantly, the determination of the total concentration of determinand in samples in which the determinand is partially complexed. For other types of analysis one of the other measurement techniques is to be preferred.

For rapid and approximate analyses, variations in S may be ignored and a plot of $[X]_A$ against ΔE (derived from Eqs (9.1, 9.2, 9.3 or 9.4), as appropriate, based on an experimental or theoretical value of S) used to derive the results. Advantages of the technique over other techniques are that electrode calibration is unnecessary, only one standard solution is required and calibration drift is unimportant; however, two potential measurements per sample are necessary. Hulanicki, Lewandowski and Maj[9] showed that for the analysis of nitrate in tap-water (an analysis in which all the determinand is uncomplexed) even if the full Eq. (9.1) is used the accuracy and precision of a standard addition procedure are worse than those of a direct method involving a calibration curve. Thomas and Booth[10] found that in the analysis of river water with an ammonia probe the standard addition and direct methods gave comparable results.

In some samples a significant proportion of the determinand is complexed and cannot be decomplexed for direct measurement, either because no suitable decomplexing reagent exists or because decomplexation would release interferents in addition to the determinand. For the analysis of such samples a standard addition method is the most satisfactory. An example of an application involving this technique is the determination of sulfur in petroleum by means of a sulfur dioxide probe as reported by Krueger.[11] The sulfur dioxide produced by combustion of the petroleum is collected in a solution containing a large excess of tetrachloromercurate ions at pH 6.9 which complex nearly all the gas as disulfitomercurate ions. After the collection period the pH of the solution is altered to 1.2, which still leaves about 20% of the sulfur dioxide complexed: the total sulfur dioxide concentration is then determined by a standard addition method.

A standard subtraction method is used in place of standard addition in circumstances where it is not possible to prepare a standard solution of the determinand for addition to the sample, either because no sufficiently pure reagents containing the determinand are available or because the determinand solutions are unstable.

A further variation on the theme of addition techniques is that known as analate addition, described by Durst.[8] In this technique the sample is added to the standard, instead of vice versa. The prime advantage of this technique is that it may be used for the analysis of volumes of sample so small that they cannot be measured by any other method because there is insufficient to cover the ends of the electrodes. The method can, however, only be used for samples in which the determinand is completely uncomplexed.

Techniques involving multiple standard additions are dealt with in the next section.

Gran's plots

Gran's plots were devised by Gran[12] in 1952 as a way of linearizing the data obtained in potentiometric titrations and thus easily and precisely locating the equivalence points of titrations. The plots are used in work with ion-selective electrodes and gas-sensing probes both for this original purpose and also for linearizing data from multiple standard addition procedures. The technique may be applied to both complexometric and precipitation titrations.

The theory associated with these plots is straightforward. The response of an

ion-selective electrode to a monovalent cation, X, in solutions free from interferents, may be represented by the Nernst equation:

$$E = E° + S \log_{10} a_X \qquad (9.5)$$

where S is the slope of the electrode response. Rearrangement of Eq. (9.5) gives

$$\text{antilog}\,(E/S) = \text{constant} \times a_X$$

where the constant is antilog $(E°/S)$. Thus, antilog (E/S) is proportional to a_X
and may be plotted, as a measure of a_X (along the ordinate) against the volume
of titrant added to a titration (along the abscissa) to give a linear plot. This is a
Gran's plot. A particularly valuable feature of the plot is that it allows the line
obtained from titration data to be extrapolated back to an intercept on the
volume of titrant axis at $a_X = 0$: thus, the equivalence points of potentiometric
titrations may be obtained.

In a conventional potentiometric titration, in which a titrand X is titrated
with a titrant Y and the course of the titration is followed by an ion-selective
electrode which senses X, the shape of the titration curve will be sigmoid as
shown in Fig. 9.2 (curve A). In some cases a single electrode may sense both X
and Y. If, for example, X is Br^- and Y is Ag^+, a bromide electrode would sense
the decrease in bromide ion activity before the equivalence point and the increase
in silver ion activity afterwards; the curve should then be symmetrical about the
end-point. The shape of the curve is less symmetrical if the stoichiometry of the

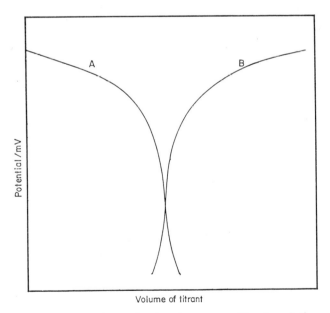

Fig. 9.2   Potentiometric titration curves (for description
see text)

reaction is other than 1:1. Similar curves are also obtained if the electrode used only responds to the titrant, as, for example, when a sulfate sample is titrated with lead ions with the lead electrode used as sensor: the curve shape obtained is shown in Fig. 9.2 (curve B).

The curves of Fig. 9.2 may be transformed into the straight lines in Fig. 9.3

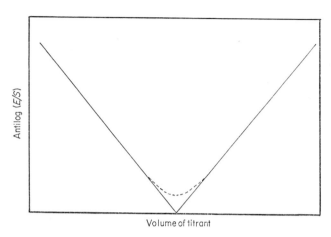

Fig. 9.3   The transformation by means of Gran's plot of the curves in Fig. 9.2 (for explanation of dotted line see text)

by means of Gran's plot. Instead of E, values of antilog (E/S), which are directly proportional to the activities (and, it is assumed, the concentrations) of the sensed species at each point (Eq. 9.6), are plotted along the ordinate against the volume of added titrant. Naturally the decrease in titrand concentration is proportional to the volume of added titrant and the resulting plot is linear. The equivalence point may then be formed by extrapolating the line to find the volume of titrant which would have to be added to bring the determinand concentration to zero—or antilog (E/S) to zero.

In practice the curves always depart quite a lot from ideality, depending on the reaction involved, because the concentration of X is never reduced quite to zero by reaction with Y. In the example of the titration of Cl^- by Ag^+ followed by the silver/sulfide electrode, the silver ion concentration, $[Ag^+]$, always has a value given by

$$[Ag^+] = K_{so}(AgCl)[Cl^-] \tag{9.7}$$

where $K_{so}(AgCl)$ is the solubility product of silver chloride. At the equivalence point $[Ag^+] = [Cl^-]$ and therefore the concentration of silver ions is not zero but

$$[Ag^+]_{eq} = K_{so}^{1/2}(AgCl) \simeq 10^{-5} \text{ mol } l^{-1} \text{ at } 25\,°C$$

The shape of the resultant Gran's plot curve near the end-point is as shown by the dotted line on Fig. 9.3. The further away the titration is from the equivalence point the less effect the appreciable solubility of the silver chloride has on the readings. Also, at the higher concentrations the electrode readings will be more stable and reproducible. Hence, by extrapolating from the straight line through the points furthest from the equivalence point, the result is more reliable than by using the values close to the equivalence point. This is an important advantage of the use of Gran's plots.[13,14]

An added difficulty in the calculations, on the experimental results to give the data for plotting, is that the concentrations of the sensed species must be corrected for the increases in the volume of the sample due to the addition of the increments of titrant: otherwise, the decrease in concentration due to the dilution by the titrant is interpreted as being caused by reaction with extra titrant. The concentrations are corrected by use of the formula

$$[X]_c = [X]_i(V_A + V_B)/V_A \qquad (9.8)$$

where $[X]_i$ and $[X]_c$ are the initial and corrected values of $[X]$, respectively. Such correction is not necessary if the titrant is generated coulometrically, so that the sample volume remains constant, or if the sensed species is buffered in the sample, so that the activity sensed by the electrode is unaffected by the dilution. An example of the latter case is the titration of cadmium ions in a citrate buffer with CDTA with the cadmium electrode used as sensor:[15] the cadmium activity is buffered by the complexation with citrate ions (primarily added to fix the ionic strength and pH). Nevertheless, in general, volume correction of the calculated concentrations—or values of antilog (E/S)—is essential.

It is clear that the calculations associated with Gran's plots are tedious, although they can be done quite rapidly with scientific calculators. By far the most convenient method of plotting Gran's plot data is to use the special Gran's plot paper (antilog-linear paper) produced by, among others, Orion Research Inc.[13] This special graph paper is available in two forms, with and without built-in correction for volume change. The more commonly used type with volume change correction automatically corrects the readings, according to Eq. (9.8), for a 10% change in sample volume during the titration (or 1% per major division of the axis); thus, if the sample volume is 100 ml, the total volume of added increments of titrant must finally be 10 ml. The ordinate of the paper is antilogarithmic, so that values of E plotted on it become converted to antilog (E/S) (the theoretical value of S is assumed), which avoids the calculation. Thus, the measured potentials are plotted directly on the ordinate against the titrant volume along the abscissa; the intercept of the extrapolated line on the abscissa (i.e. at $[X] = 0$, which is where $E = \pm\infty$, depending on whether X is an anion or a cation) marks the end-point and, hopefully, the equivalence point. Errors are introduced if the value of S is not theoretical, as is often the case; although these can be compensated,[13] the effort involved may make it simpler, especially if accuracy and precision are important, to calculate the correct antilog (E/S) values and plot them on to ordinary graph paper.

Closely allied to the use of Gran's plots to follow potentiometric titrations is their application to multiple standard addition procedures. Instead of adding just one increment of a standard solution of the determinand to the sample, as in the normal standard addition procedure, several are added. The resulting build-up of determinand concentration is entirely similar to the increase in titrant concentration after the end-point of a titration, but starting from the sample concentration instead of nominally zero. If the results are plotted on volume-corrected Gran's plot paper as before, with the electrode potential plotted up the antilogarithmic ordinate, a graph such as that illustrated in Fig. 9.4 is

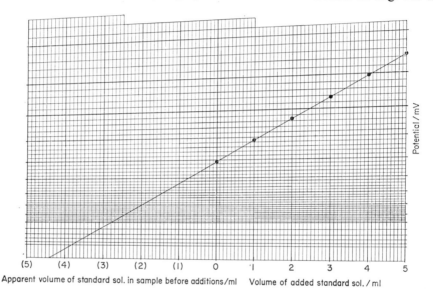

Fig. 9.4   An example of the use of Gran's plot paper (10% volume-corrected) (Orion Research Inc.) for analyses by the multiple standard addition technique

obtained. The original sample concentration may then be simply determined from the intercept of the line on the abscissa. In the example shown the original sample concentration, $[X]_A$, (assuming that the sample volume is 100 ml) is apparently equivalent to that concentration which would have been produced by the addition of 4.4 ml of the standard solution of determinand to pure sample background (as the intercept on the axis is where $[X] = 0$): thus the concentration to be measured in this case is given by

$$[X]_A = \frac{4.4 \times [X]_B}{100}$$

As with potentiometric titrations, Gran's plot paper without volume correction must be used when the standard increments are added coulometrically or the sample is buffered with respect to the species sensed by the ion-selective electrode.

There are two advantages in using this multiple standard addition technique. First, because the result is dependent upon several readings, the effect of random errors in individual readings on the result is reduced, and the accuracy and precision are greatly increased. Second, the technique may be used for the analysis of samples containing interferents at concentrations several times those which can be tolerated in direct determinations. This is because the final concentration of determinand in the sample after the additions can be made sufficiently high for the interferents to have virtually no effect on the electrode response; the result is then calculated by extrapolation from these interference-free readings. This is analogous to the technique, already described for potentiometric titrations, whereby effects due to solubility of precipitates near the equivalence point are discounted.

A variant of this multiple standard addition technique has been described by Frant, Ross and Riseman,[16] whereby, instead of using an electrode which senses the determinand directly, the determinand concentration is sensed via an intermediate reaction. To samples containing cyanide ions, they added a small concentration, typically 10^{-5} M, of dicyanoargentate(I) ions ($Ag(CN)_2^-$): this fixed the silver ion activity, which was sensed by a silver/sulfide electrode, at a value given by

$$a_{Ag^+} = a_{Ag(CN)_2^-}/\beta_2 a_{CN^-}^2 \qquad (9.9)$$

where β_2 is the stability constant for the formation of $Ag(CN)_2^-$ ions. Increments of standard cyanide solution were added to the samples and the resultant cell potentials plotted on Gran's plot paper without volume correction (because the silver ion activity is buffered by the large excess of cyanide ions). All samples were pretreated by addition of potassium hydroxide and EDTA to fix the ionic strength, to raise the pH to the range 11–12 and to decomplex cyanide from metal-cyanide complexes: complexes of Cr^{3+}, Cu^+ and Ni^{2+} were determined after initially heating the samples at 50 °C for 5 min at pH 4. The multiple standard addition plots obtained are shown in Fig. 9.5.

Instead of manually adding the increments of standard solution to the sample from a pipette or burette and plotting the results, the Gran's plot procedure may be automated by use of a modified Technicon AutoAnalyzer and the result calculated by computer.[17]

When a method involving a Gran's plot is developed, several points have to be kept in mind. The difference between the highest and lowest potentials to be plotted should be within the approximate range 20–50 mV for electrodes sensing monovalent ions or 10–25 mV for divalent ions when the special graph paper is used. At least four or five additions of standard and potential measurements are desirable for the full increase in precision and accuracy afforded by the technique to be realized. For titrations the stoichiometry of the reaction taking place must be constant throughout the titration or the plot will produce a curve; for complexometric titrations, in particular, this is sometimes difficult to achieve and may require special sample pretreatment.[15] Additionally, samples must be pretreated as for direct measurements to ensure constancy of the ionic strength, correct pH, etc.

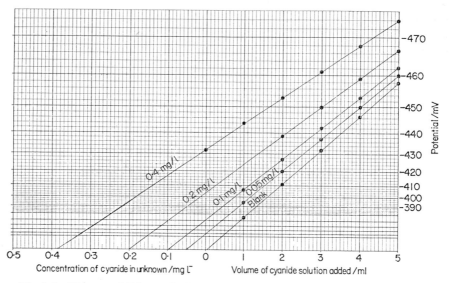

Fig. 9.5   Known additions of 0.01 mg l⁻¹ cyanide to known samples containing
cyanide in the range 0–0.4 mg l⁻¹. The results are plotted on Gran's plot paper
without volume correction. Samples were adjusted to pH 11 [16]

Potentiometric titrimetry

In a potentiometric titration the sample is titrated with a suitable titrant and the
decrease in titrand activity or increase in titrant activity is followed with an
ion-selective electrode. A curve of the shape illustrated in Fig. 9.2 is obtained,
and the end-point of the titration is assumed to be at the inflection point of the
curve. The theory of titrations is now well known and has been set out in detail
by Bishop;[18] this text gives methods for calculating the feasibility of particular
titrations. An account of potentiometric titrations has been given by Davis:[19]
although this text appeared before the widespread use of ion-selective electrodes,
it is still useful.

Potentiometric titrimetry is seldom the preferred manual technique for
performing an analysis, although it can be useful in some circumstances. Usually
when the accuracy and precision are important (the normal reasons for resorting
to titrimetry in place of direct methods), the analysis may be done more rapidly
and the end-point located more precisely by a Gran's plot method, using the
special graph paper for plotting the graph. However, potentiometric titrations,
unlike Gran's plot methods, may readily be automated by use of any of the
commercially available automatic titrators; this is particularly useful in quality
control and similar routine applications. Also, for the analysis of occasional or
unique samples for which no Gran's plot method has been developed, it is much
simpler to use a potentiometric titration, since the development of a Gran's plot
method is not always straightforward, especially for complexometric titrations.

One of the major advantages of the titration techniques, both potentiometric titrimetry and Gran's plot methods, is that they enable a large range of species, not directly sensed by ion-selective electrodes or gas-sensing probes, to be measured. Thus, for example, metals such as zirconium, thorium, lanthanum, samarium,[20] nickel, zinc and cadmium[21] may all be titrated with EDTA with use of a cupric ion-selective electrode as sensor and cupric ions as indicator (via the activity of 'free' cupric ions in equilibrium with the Cu–EDTA complex). In general, the relative standard deviation for these methods is slightly less than 1%[20] for metal ion concentrations between 10^{-3} and 10^{-4} M, which is considerably worse than is obtainable in manual titrations using visual indicators such as xylenol orange: however, the visual methods cannot be used at all in highly coloured or turbid samples.

A novel technique for performing semi-automatic potentiometric titrations has been proposed by Fleet and Ho;[22] they have termed it 'gradient titration'. In this technique, which involves the continuous-flow system diagrammatically represented in Fig. 9.6, a continuous flow of pretreated sample is mixed with a steadily increasing concentration of titrant. The concentration of titrant in the container marked B, which initially contains only distilled water, increases linearly with time at a rate dependent on the concentration of the standard titrant solution in A, the volume of solution in B and the pumping rates of solution to and from B. The sample stream plus titrant is mixed and then passes the electrode. In the example shown, in which a stream of the sulfide sample is mixed with a steadily increasing concentration of mercuric ions, the ion-selective electrode used is a sulfide electrode with the sensing pellet drilled so that the sample passes through it. When the electrode eventually reaches the predetermined potential at which the sulfide ion activity in the sample stream has

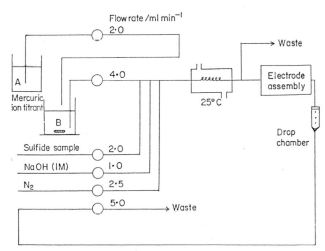

Fig. 9.6   Flow diagram for the gradient titration of sulfide
samples with mercuric ions[22]

been reduced to the end-point value, the mercuric ion concentration at that time is equivalent to the required sulfide ion concentration. This concentration of mercuric ions at the end-point is calculated from the known concentration gradient and the titration time required, from the start of the titration, to reach the end-point detected by the electrode.The relative standard deviation obtained in the determination of ten replicate sulfide samples of concentration 4×10^{-4} M was 1.16%. With some further modification the apparatus could be made fully automatic and used for many titrations. Advantages of the technique are that automatic burettes are not needed and that the titre may be determined very precisely by time measurements. These features have been common for several years in commercially available automatic titrators using conventional titration of the sample with a standard solution delivered by a peristaltic pump; for example, the E.I.L. Model 81 titrators are based on this system, although the titre is measured by counting the revolutions of the pump required to reach the end-point. However, an added advantage of the gradient titration technique is that, by using a flow-through electrode, the response speed of the electrode is optimized, which allows very sharp end-points to be obtained rapidly. On the debit side, the system relies critically on the maintenance of precisely known pumping rates and ratios which are inevitably difficult to achieve over a long period of time; in particular, the ratio of the flow of solution into and out of vessel B must be kept as close as possible to 1:2 or the concentration of titrant in B will no longer be a linear function of time.

The range of applications of ion-selective electrodes in potentiometric titrimetry is limited by the response times of the electrodes. In the conventional configuration, in which the electrodes are immersed in a stirred sample in a beaker and the titrant is added from a burette, the response time of many of the electrodes, particularly those based on organic ion exchangers or neutral carriers, may well be at least 15 s at the beginning of the titration but several minutes at the lower concentrations near the end-point. As, for the more precise location of the end-point, it is customary to take several readings in this region, the plotting of the curve may therefore be an unacceptably protracted business. The inorganic salt-based electrodes and glass electrodes are usually more satisfactory than the others, but even so the titrations take much longer than with visual indicators. In general, the response times of the electrodes are shorter at higher concentrations of determinand or titrant and therefore the titrations become more viable for stronger samples; it is not possible, however, to give any limits below which potentiometric titrations cease to be feasible, because there is such a large difference between the performances of the different electrodes. Where problems of response speed arise, the Gran's plot technique is again much more satisfactory, as the result may be calculated from fewer readings, which are taken well away from the equivalence point region.

Errors in potentiometric titrations, due to the presence of species in the sample interfering with the response of the electrode to the determinand or titrant, have been treated theoretically for both precipitation[23] and complexometric[24] titrations. The distortion of the curve caused by the constant response to the interferent produces separation of the end-point from the equivalence point. This problem may be overcome, if the interferent is present in relatively moderate amounts, by resorting to Gran's plot methods again; Gran's plot methods will also give more accurate results than potentiometric titrations in double titrations, such as the successive titration of bromide and chloride ions in a sample with silver nitrate solution.[14] Despite such errors and complications, both potentiometric titrations and Gran's plot methods are nearly always several times more accurate and precise than direct methods of analysis when comparable methods are available, although the latter are always much more rapid.

Thus, several techniques are available for the analyst wishing to employ ion-selective electrodes or gas-sensing probes. The choice between them is usually made by finding where the balance lies between speed and convenience on the one hand and the required accuracy and precision on the other.

METHOD DEVELOPMENT

Once the choice of technique is made, a method is then developed for the analysis of the samples, during which the following factors, which are not in order of importance, must be considered: (i) concentration range of determinand in samples, (ii) choice of electrodes or probe, (iii) sample pH, (iv) sample ionic strength or osmotic strength, (v) state of the determinand in the sample, (vi) interferents in the sample, (vii) standardization of the cell, (viii) sample velocity at electrode/probe sensing surface, (ix) sampling and the stability of samples, (x) sample temperature and (xi) sample pressure. Although this list does not cover all eventualities, all major points are included: many of the points are clearly interdependent. It may be noted that several factors do not appear which would certainly appear in a similar list prepared for development of spectrophotometric and other techniques; the most important of these are sample colour and turbidity, which have no direct effect on ion-selective electrode measurements.

Each of the points in the list will be discussed in turn and its influence in ion-selective electrode techniques compared, where appropriate, with its influence in other analytical techniques.

(i) The concentration range of the determinand in the samples to be analysed naturally has a major effect on the method developed. One of the important advantages of both ion-selective electrode and gas-sensing probe methods over other analytical techniques is that they can measure determinands directly without modification over several orders of magnitude of concentration. However, the electrodes are less suitable for the precise determination of determinands in a narrow concentration range: where possible in such cases it is better to use an analytical method which gives a linear response to concentration, such as titrimetry or spectrometry. For high concentrations of determinand, direct ion-selective electrode methods are much less accurate and precise than titrimetric methods, but as the concentrations decrease, titrations rapidly become less and less satisfactory, and are useless in most cases below 10^{-4} M determinand. The accuracy and precision of direct ion-selective electrode/gas-sensing probe methods, however, remain virtually unaltered down to the limit of Nernstian response of the sensor; with care an accuracy of better than $\pm 1\%$ is attainable throughout the range, with a relative standard deviation also of about 1%. For determinations of samples containing the determinand at a concentration between the limit of Nernstian response and the limit of detection of the sensor, it is wisest to seek an alternative method of analysis where possible. Thus, for example, for samples containing 10^{-5} to 10^{-6} M calcium ions a technique such as flame-emission spectrophotometry should be considered instead of an electrode method.

(ii) The choice of the most suitable ion-selective electrode and reference electrode or gas-sensing probe from the many which are available is, of course, extremely important. The factors involved in the selection of the reference electrode and the bridge solution were considered in Chapter 2; in summary, the reference electrode should give a stable potential and the choice of bridge solution should ensure minimum contamination of the sample and small and stable junction potentials. The criteria for selection of the most suitable electrode or probe were elaborated at the end of Chapter 3.

(iii) Nearly all samples need pH adjustment before measurement; this is achieved by the addition of a buffer, which usually also serves to fix the ionic strength of the samples. The final pH to which the samples are adjusted is chosen such that it is within the optimum working pH range of the sensor, all the determinand, or a constant proportion of it, is in the form which is sensed and interferences are (whenever possible) masked or precipitated. As already noted in earlier chapters, the optimum working pH range is concentration-dependent and usually becomes narrower as the limit of Nernstian response is approached. It is wise to check with a pH electrode that the desired pH is actually achieved after addition of the buffer for all possible sample compositions, since the buffer capacity of samples may vary widely. Thus, for example, in the determination of sulfur dioxide in fruit juices with a sulfur dioxide probe it is found that the buffer capacities of such juices as orange and lime differ considerably and the pH at which they are buffered ranges from 2 to 5; it is therefore essential, if a uniform method is to be used for all types of juice, that sufficient acid be added for the pH of even the most strongly buffered juice to be reduced to the required value ($\leqslant 0.7$).

The optimum pH ranges for samples for measurement with the different electrodes and probes are listed in the relevant chapters.

(iv) As discussed in Chapter 3, if the concentration (rather than the activity) of a determinand is being measured with an ion-selective electrode, the ionic strengths of both samples and standards must be held constant by the addition of an electrolyte which will not interfere with the electrode response. This fixes the activity coefficient of the determinand, thus ensuring that the concentration which it is wished to determine is directly proportional to the determinand activity, which the electrode actually senses.

For gas-sensing probes the situation is slightly more complicated since two factors are involved. First, it is advisable to fix the composition of the sample and standards as far as possible so that the Henry's law constant remains, in fact, a constant; however, the Henry's law constant is not very sensitive to small changes of sample background, and the buffer added to produce the correct pH for measurement usually deals with this aspect adequately. Second, and more important, for gas-sensing probes with microporous membranes or 'air-gaps', it is necessary to match the total concentration of dissolved species in the sample to that of the internal electrolyte, to prevent the osmotic drift caused by transfer of water vapour (see Chapter 7). This matching can usually also be achieved by appropriate choice of buffer; however, for some samples, such as Kjeldahl digests to be analysed with an ammonia probe, modification of the internal electrolyte of the probe may be necessary.

These problems arise from the fact that the ion-selective electrodes directly sense ionic activities and the gas-sensing probes sense partial pressures of dissolved gases, whereas the analyst nearly always wants to measure ionic concentrations. Most other techniques, such as AAS, measure concentrations directly and such problems do not arise.

(v) In many samples much of the determinand is bound up in inert or labile complexes of various stabilities; for example, in natural waters a significant proportion of the fluoride ions present is complexed by aluminium and ferric ions. The uncomplexed activity of determinand ions in equilibrium with these complexes may be measured directly with an ion-selective electrode, provided that the electrode will operate in the samples without sample pretreatment. Addition of pH buffer, etc., will nearly always disturb the equilibria between the determinand and its complexes. However, if, as is usually the case, the parameter to be measured is the total concentration of determinand, then a reagent must be added to the samples to mask the complexing species. In the given example of fluoride determination, CDTA is added (see Chapter 5) which itself complexes the aluminium and ferric ions, thus freeing the fluoride ions for measurement. These fluoride complexes are broken down relatively easily, but some complexes are more difficult to treat and a more elaborate pretreatment is necessary to decomplex all the determinand. For total cyanide determination in samples containing the highly stable complexes of Cr^{3+}, Cu^+ and Ni^{2+} a two-stage treatment is necessary,[16] as mentioned previously in this chapter. Similarly, for the determination of 'total' sulfur dioxide in glucose syrups a two-stage procedure is necessary, involving a preliminary alkaline pretreatment to break down the bisulfite addition compounds before the final acidification. Several other examples of similar sample pretreatments appear in the Applications sections of Chapters 5 and 6.

In comparison with other techniques, ion-selective electrode and gas-sensing probe techniques are much more flexible in respect of the choice of the state of the determinand which it is possible to measure. Direct methods offer the possibility of measuring the 'free' determinand activity without disturbing the complexation equilibria or, alternatively, of measuring the 'total' activity after adjusting the equilibria in a controlled way. In contrast, in colorimetric methods the amount of determinand measured depends on the relative strengths of the complexes of the determinand with the sample constituents and with the chromogenic reagent, coupled with the rates of the reactions of the various species with each other and the length of the colour development period: the situation is thus quite complicated. In AAS it is much clearer, since the parameter measured is the total concentration of determinand species entering the flame (excluding side reactions in the flame).

(vi) It is very important in the development of a method to assess the maximum possible concentrations of all interferents which are at all likely to be present in samples. Then the amount of interference, if any, which could arise may be calculated in the manner described in Chapter 3 and steps may be taken, where necessary, to mask, precipitate out, destroy or otherwise render innocuous the interferents. It is essential in all cases to make the estimate with the maximum possible concentration of interferent being used rather than the average concentration, if any confidence is to be justifiably placed in the results. Most of the inorganic salt-based electrodes and several of the electrodes based on organic ion exchangers or neutral carriers, obey quite closely the equation

$$E = E^\circ \pm \frac{2.303RT}{n_A F} \log_{10} \left(a_A + \sum_B k_{AB} a_B^{n_a/n_B} \right) \qquad (3.10)$$

Thus, true readings of the concentration of the determinand are indistinguishable from erroneous readings bolstered up by the response to the interferents. Only when the interference is so gross that it causes potential drift is any fault apparent.

If the interferent rapidly attacks the electrode or otherwise irreversibly impairs the response, then the interferent must be removed from the sample and never allowed to come into contact with the electrode. Thus, chloride samples containing traces of sulfide must be treated to remove the sulfide completely (usually by a two-stage procedure involving addition of a strong oxidant and removal of the excess) before the samples are presented to the probe; otherwise, for silver chloride-based chloride electrodes, an unresponsive coating of silver sulfide would be formed on the electrode surface, which could only be removed by abrasion. However, in other less drastic cases in which the interferent attacks the electrode only slowly, or not at all, it is sufficient to reduce the concentration of interferent to a level at which it will not interfere. For the analysis of nitrate samples containing chloride ions, the chloride ion concentration is reduced by precipitation with the silver ions added in the ionic strength adjustment buffer (see Chapter 6); thus, the pretreatment is performed in a single, straightforward step.

It is usually desirable to aim to develop a method with a single pretreatment stage, as such procedures are much more convenient in practice than multi-stage procedures. However, this places several constraints on the choice of pretreatment reagent. In particular, the reagent must act to remove or reduce the concentration of the interferent, when used within the normal working pH range of the electrode, and must selectively act on the interferent without affecting the determinand (although occasionally a constant level of complexation of the determinand is acceptable). It must also be ensured that the reagent is successful in removing the interferent at all temperatures at which samples are to be measured. For example, Eckfeldt[25] has shown that errors can arise, in the determination of low concentrations of sodium ions (about 1 μg Na$^+$ per litre) with a sodium ion-selective glass electrode, due to change in the ability of the buffer to remove hydrogen ion interference as the temperature changes. If a 0.02 M solution of the buffer used (diisopropylamine) is heated from 25 °C to 60 °C, the pH of the solution falls by about one unit and hydrogen ion interference increases tenfold: thus, as the temperature of the sample increases, either the concentration of the buffer added must be increased, to ensure that the correct pH is maintained for insignificant hydrogen ion interference, or, preferably, the samples should be cooled before measurement.

The form of Eq. (3.10) suggests a technique which can be used on occasion to reduce interference by monovalent ions, B, on divalent ion-selective electrodes (i.e. $n_A = 2$ and $n_B = 1$). For example, for a sample containing calcium ions as determinand at a concentration of 10^{-2}–10^{-3} M and just sufficient sodium ions to interfere, the interference may be avoided by diluting the sample. If the sample is diluted tenfold, the determinand concentration will still be well above the limit of Nernstian response of the sensor although the a_A term is now $0.1a_A$; however, the interference term is now reduced, because $n_A/n_B = 2$, to $0.01k_{AB}a_B{}^2$, and thus the amount of interference is reduced tenfold.

In (iii)–(vi) those requirements of methods which may be satisfied by the addition of a pretreatment reagent have been outlined. In addition to the functions of the reagent in these various ways, other aspects of the addition and action of the reagent may be considered.

As mentioned in (iii), the reagent chosen should not interfere with the electrode response; in addition to the need to eliminate interferents from the reagents, this also implies that the reagent should not be contaminated by the determinand. When the electrodes or probes are used in trace determinations, this is particularly important, as has been discussed in previous chapters in relation to the measurement of low concentrations of sodium, fluoride and ammonium ions.

For the calibration of the cell to remain valid, the volume ratio of sample to added reagent must remain constant. Thus, it is convenient to establish the method so that fixed volumes of sample are treated with fixed volumes of reagent: this condition is automatically satisfied in continuous-flow systems.

In some cases, by careful choice of reagents and standards, a calibration graph may be used for two methods involving different pretreatments. For example, in the determination of 'free' and 'total' sulfur dioxide in lemon squash, with a sulfur dioxide probe with a homogeneous membrane, a considerable amount of time may be saved by adopting a combined method for determining both states of the determinand sequentially. It is found experimentally that 60 ml of lemon squash requires treatment with 8 ml of 4 M sodium hydroxide (to raise the pH for decomposition of the aldehyde–bisulfite addition compounds) and then 12 ml of 4 M sulfuric acid to lower the pH to $\leqslant 0.7$ for measurement of 'total' sulfur dioxide; however, 'free' sulfur dioxide may be measured simply after addition of 5 ml of 4 M sulfuric acid to a 60 ml sample. Hence, the samples for 'total' measurements should be treated as above, and the samples for 'free' measurements and standards should be treated by addition of 5 ml of 4 M sulfuric acid plus 15 ml of water to produce the same total sample dilution (this assumes, as is normally valid, that the variations in Henry's law constant between the different treated samples and standards are negligible). Thus, the same set of calibration standards may be used for both determinations.

(vii) Standardization or calibration of the cell which is set up has already been discussed earlier in this chapter and in Chapter 3. Careful selection of the standards to use is crucial in method development, since the accuracy of the method is inevitably limited by the accuracy of the standards. Thus, standards should be carefully prepared, and formulated, so that their composition is as close as possible to the composition of the samples.

(viii) The velocity at which the solution flows over the surface of an electrode has an important effect on the value and reproducibility of the electrode potential, and on the response speed of the electrode, and can also affect the liquid junction potential of the reference electrode: reference was made in Chapter 6 to the effects of stirring on the response of the calcium and potassium electrodes. It is therefore necessary that the velocity should be both adequately high and constant.

In the manual methods, in which the samples are contained in beakers, the rate at which the solution flows across the electrode surface, and past the liquid junction of the reference electrode, is controlled by the rate of stirring of the sample and the disposition of the electrodes in the beaker. Therefore, these two parameters should be as constant as possible: the stirring should be brisk but not so fast that a vortex is created. The two conditions are conveniently realized if the electrodes are used in a continuous-flow system: for this purpose the electrodes may be fitted into a special flow cell, fitted with a flow-through cap, or for home-made electrodes with solid state, non-glass membranes, the membranes may be drilled so that the sample flows through them (see, for example, Fig. 9.8 in which possible configurations are shown). Eckfeldt and Proctor[26] demonstrated the important effect that the design of a flow cell has on the measured potentials. In the measurement of low concentrations of sodium ions (about $0.1\ \mu g\ Na^{+}$ per litre) with a sodium electrode, they obtained differences of more than 10 mV between the electrode responses in a solution passing the electrode at constant flow rate but at different flow velocities (produced by changing the volume of that part of the flow cell round the electrode bulb). The most stable and sensitive response was obtained at the highest sample velocity.

If the flow rate of the solution is high, however, it is essential to guard against the development of streaming potentials, which can lead to large errors.[27] These potentials can produce oscillation in the response of the cell, at a frequency equal to that at which the bar or roller of the peristaltic pump producing the flow passes over the pumping tube (the pulsing frequency), and also shift of the baseline. To minimize these effects, the conductivity of the sample should be high (often achieved by the pretreatment reagent) and the flow tubing used should not be of unnecessarily narrow bore. Also, a high pulsing frequency, produced by using a peristaltic pump with many rollers, and the use of mechanical pulse suppressors or of electronic pulse suppression help to produce a more satisfactory cell response.[27]

Pulsing of the sample flow in continuous-flow systems incorporating gas-sensing membrane probes can also create problems, but of a different kind. The pulses cause variations in the pressure inside the flow cell or flow-through cap, which in turn produce membrane movement and, hence, oscillation of the thickness of the thin film of electrolyte between the membrane and the internal pH electrode. As the membrane moves away from the pH electrode, some of the internal electrolyte from the bulk of the probe will be drawn into this thin film: this electrolyte, which will not be in equilibrium with the sample, will cause the apparent concentration of determinand to drop. The apparent concentration will reach a minimum, as the thin film reaches maximum thickness, and will then increase as determinand flows through the membrane to approach chemical equilibrium with the new thin film, which meanwhile contracts. The resultant oscillation, in the potential output of the probe, can be entirely obviated by minimizing the intensity of the pulses and increasing the tension of the pH electrode against the membrane.

Despite the need to consider the effect of sample velocity carefully, it seldom limits the application of electrodes or probes, and for the analysis of samples in beakers containing the determinand at concentrations higher than approximately 10^{-4} M errors from this cause are uncommon. If variations in the stirring speed do produce substantial potential changes, then the first suspicion should be that the junction potential is varying: such variation could be caused by either the liquid junction of the reference electrode becoming blocked or by the bridge solution of the reference electrode not being sufficiently equitransferent.

Such problems with sample velocity are not encountered in many other analytical techniques (such as titrimetry with visual indicators or AAS) but do arise in techniques such as polarography.

(ix) Sampling and the stability of samples must be prime considerations in the development of all analytical methods for whatever technique. The way in which the sample is abstracted for analysis often has more effect on the eventual answer than any other factor; thus, for example, if the nitrate content of a river is to be determined, it is important to decide whether the required sample is of water near the surface or near the bed of the river, in the centre or near the banks. After the sample is taken, it is then stored in an inert container which may be easily cleaned, such as a polythene bottle, and any necessary preservative added to prevent sample deterioration: in the case of nitrate samples, a mercuric salt may be added and the samples cooled to 5 °C to reduce biological consumption of the nitrate. In general, samples containing less than 10^{-4} M should be stored in inert plastics containers instead of in glass. A detailed discussion of sampling and how to deal with unstable samples in the context of the analysis of waters is given in the book by Wilson.[28]

Occasionally samples are so unstable that special precautions must be taken to ensure that the sample composition does not change during the actual analysis. For example, sulfide samples are so prone to oxidation that in a direct manual method of analysis with a sulfide electrode all that is recorded is a drifting response unless the sample is stabilized. As mentioned in Chapter 5, the stabilization may be accomplished by the addition of an anti-oxidant buffer containing a reductant such as ascorbic acid. A somewhat different situation arises in the determination of sulfur dioxide with a sulfur dioxide probe: the sulfur dioxide which is generated after the necessary acidification of the samples is so volatile that measurements must be taken in a closed cell or flask (see Fig. 7.4).

(x) The performance of the various cells employing ion-selective electrodes or gas-sensing probes depends a lot on the sample temperature, as has been described several times in different chapters. The temperature affects the cell in several ways, of which the most important are that it alters the standard potential ($E°$) of the ion-selective electrode, the response slope (mV decade^{-1} activity change), the potential of the reference electrode, the activity of the determinand (by alteration of the activity coefficient, γ) and also, sometimes, the degree of complexation of the determinand and the efficiency of the pretreatment reagent in removing interferents (see vi).

Very few data have been published on the performance of any of the electrodes or probes above 40 °C; therefore it is advisable, especially with applications involving the electrodes based on organic ion exchangers or neutral carriers, to cool all samples to either room temperature or to a fixed 25 °C before measurement whenever possible. For the greatest precision both samples and standards should be thermostatted to the same temperature; however, the temperature coefficients of many cells are low enough for satisfactory results to be obtained by taking all results at room temperatures.

As discussed in Chapter 7, the gas-sensing probes with microporous membranes or 'air-gaps' suffer severe hysteresis after change of sample temperature owing to osmotic effects; hence, it is particularly important to ensure that the temperatures of all solutions, whether samples or standards, are identical when presented to the probe. Use of a thermostatted continuous-flow system is thus advantageous with gas-sensing probes.

Most cells do not respond in a sufficiently theoretical manner for automatic temperature compensation to be used in the manner commonly practised in pH measurement. Another drawback is the paucity of published isopotential data, the only compilation being that of Negus and Light.[29] Moreover, many pH meters for use with ion-selective electrodes or gas-sensing probes are not capable of this compensation on the expanded scale ranges.

Thus, it is clear that substantial fluctuations in sample temperature (> about 10 °C) can cause serious problems in applications of ion-selective electrodes and gas-sensing probes. Such difficulties are also likely to arise in, for example, spectrophotometric and polarographic methods but would be avoided in visual titrimetric and AAS methods.

(xi) The effect of sample pressure on the cell will normally only need to be considered in on-line industrial monitors. If the sample pressure rises above atmospheric pressure, the reference electrode should be pressurized, conveniently by the addition of an external reservoir, so that a flow of bridge solution out through the liquid junction is always maintained and the sample is prevented from getting into the reference electrode and contaminating it.

Gas-sensing probes are completely unsuitable for measurements at varying pressure, but could probably be used to analyse high-pressure samples if the pressure inside the probe were to be equalized with that outside.

The development of analytical methods employing ion-selective electrodes or gas-sensing probes is not usually at all complicated, despite the several points which have been discussed. Indeed one of the prime factors leading to the rapid expansion in the use of the sensors has been the ease with which methods may be developed, for a wide range of determinands in many different types of sample, without the need for protracted research or development work. Because of the simplicity of direct, manual methods and perhaps also because of the analogy with pH measurement, it is only more recently that the electrodes and probes have become at all widely used in continuous-flow analysis systems in the laboratory; however, these systems are attractive, since they offer the opportunity for much greater analytical precision. These laboratory systems have been preceded by several years by industrial monitors incorporating ion-selective electrodes or gas-sensing probes; the most popular of these industrial monitors are those to measure sodium, ammonium, fluoride and nitrate ions.

CONTINUOUS-FLOW SYSTEMS—LABORATORY AND INDUSTRIAL

Automatic or semi-automatic analysers based on continuous-flow systems have several practical advantages, other than the obvious ones of speed and convenience, which lead to a higher precision than is readily feasible with direct manual methods. These advantages arise from the standardized way in which the treated sample is presented to the sensor for measurement. In particular, the ratio of sample to reagent, the sample velocity across the membrane surface and the temperatures of both the sensor environment and the sample are held constant. In addition, in systems fitted with sampler units for dispensing discrete samples, the times during which the electrode or probe is exposed to the sample and wash solution are held constant.

A typical flow system for use with electrodes or probes is shown in Fig. 9.7;[30] in this case the sensor is an ammonia probe fitted with a flow-through cap. The ammonium samples are pumped from the sampler by the four-channel peristaltic pump and segmented by air, to improve separation of samples and mixing with the buffer. The segmented samples then meet the flow of buffer

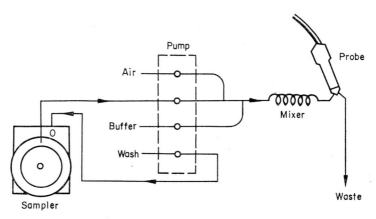

Fig. 9.7   Diagram of flow system[30]

(sodium hydroxide solution in this case), pass through the mixing coil, which ensures that the segments of treated sample are thoroughly mixed, and then flow straight past the membrane of the ammonia probe and out to waste. In systems employing other sensors it is sometimes necessary to pass the sample through a de-bubbler before it reaches the sensor. This prevents erratic response due to interference with the surface equilibria, and partial disruption of the electrical contact through the solution to the reference solution, as the air bubbles pass over the membrane. However, when the ammonia probe is the sensor, there is no external reference electrode, and the air bubbles collect ammonia from the treated samples during the mixing stage so that the partial pressure of ammonia 'seen' by the probe does not vary as the bubbles pass the membrane. The wash solution chosen has a concentration of about half that of the weakest sample expected: this is preferable to using a distilled water wash, since it raises the measurement baseline and effectively shortens the response time of the system. The buffer is pumped through continuously: it is important to ensure that the supply of wash solution or sample does not fail, as exposure of the probe to undiluted buffer would rapidly cause osmotic drift. Furthermore, when other sensors are used, particularly the electrodes based on organic ion exchangers or neutral carriers, it is found that exposure to the undiluted buffer can produce hysteresis. The response of the system illustrated to standard solutions of ammonium ions is shown in Fig. 9.8:[30] the standards were run at the rate of 60 per hour. Buckee,[31] who used a similar system but thermostatted the probe by immersing it up to the cap in a water bath to minimize baseline drift, measured total nitrogen by a Kjeldahl method, using the probe to measure the ammonium ion concentration in the digests, at the rate of 60 per hour with a relative standard deviation of approximately 0.5%. Another example of the use of a laboratory flow system is shown in Fig. 8.2.

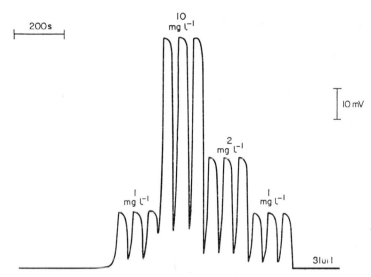

Fig. 9.8 Response of an ammonia probe in the flow system.[30]
Buffer: 0.7 M sodium hydroxide. Wash solution: 0.5 mg l^{-1} NH_4^+.
Sampling time: 40 s. Wash time: 20 s. Pumping rates: sample, 4.1 ml min^{-1}; buffer, 0.7 ml min^{-1}; air, 2.0 ml min^{-1}; wash, 4.1 ml min^{-1}

Some configurations of flow-through cell and sensor end-caps which can be used in flow systems are depicted in Fig. 9.9.

The specification of an industrial analyser differs from that of the laboratory continuous-flow system in several respects. In particular, the industrial analyser should be designed so that it may be left running for long periods unattended, and thus must be extremely reliable and must consume only small quantities of reagent (up to about 5 litres per week is usually acceptable). An industrial analyser is usually required to monitor dirtier samples than are commonly encountered in the laboratory; thus, if the analyser is to continue working over long periods without clogging of the tubes or coating of the sensor or reference electrode, it is essential to ensure adequate filtration of the sample. To produce a filtration system which will work satisfactorily for long periods is perhaps the most difficult part of the design of the monitor.

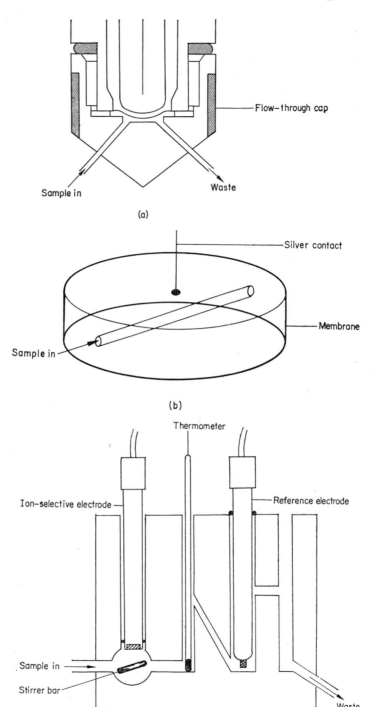

(a)

(b)

(c)

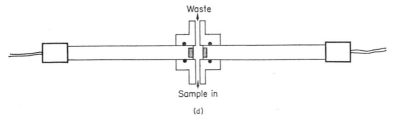

Fig. 9.9 Various flow-through sensors and cells: (a) flow-through cap fitted to a gas-sensing membrane probe; (b) inorganic salt-based membrane drilled for sample flow through the membrane; (c) typical flow cell in an industrial monitor, using an ion-selective electrode and a reference electrode downstream; (d) a flow system using two ion-selective electrodes to make a cell without liquid junction

Few industrial samples can be analysed sufficiently accurately or precisely by direct immersion of the electrodes in the sample stream, although for pH measurement this is usually possible. With pH electrodes the membrane may be kept clean automatically by mechanical wiping or ultrasonic agitation; however, the membranes of most of the other sensors are not durable enough to permit this and would therefore rapidly become coated with particulate matter, oil or biological growths. Furthermore, the measurement of pH is usually only required to an accuracy of ± 0.2 or, sometimes, ± 0.1 pH units (± 12 mV or ± 6 mV); however, when converted into concentrations these uncertainties in potential become $\pm 58\%$ or $\pm 26\%$ for monovalent determinands ($\pm 151\%$ or $\pm 58\%$ for divalent determinands). Such large uncertainties are not acceptable in the monitoring of most ions measured with ion-selective electrodes; even if only $\pm 10\%$ is required, this corresponds to ± 0.04 pX units or ± 2.4 mV (for monovalent determinands), which is very difficult to achieve if the sample and environment temperature are liable to change and if the sample composition is not stabilized by addition of a pretreatment reagent. Another problem with the sensors is that they require fairly frequent standardization because of E° drift (approximately every 6–12 h). If they are immersed directly in the bulk sample, this can only be done after removal of the sensor from the system; then the system is no longer fully automatic but requires attention every few hours.

A more reliable way of continuously monitoring the concentration of a determinand, which gives more accurate and precise results, is to bleed off a sample from the source (process stream, effluent, river, water supply or whatever) and pump it through a filter into a thermostatted monitor. The filtered sample stream may then be pretreated by an appropriate reagent, mixed to ensure homogeneity, and presented to the probe in a closely controlled way. In some commercial monitors the sample stream may be switched off at a preset frequency (typically every 6 or 12 h) and replaced by a stream of a standard solution from a reservoir in the monitor; the calibration point of the monitor is then reset electronically after the potential of the cell has reached equilibrium. The concentration of the standard solution should be at approximately the middle of the range of the instrument. With such an instrument it is possible to analyse continuously a determinand with an accuracy and precision of $\pm 5\%$. A diagram of a monitor which incorporates all these features is given in Fig. 9.10. The constant head unit on the sample inlet ensures that the pressure at this point remains fixed.

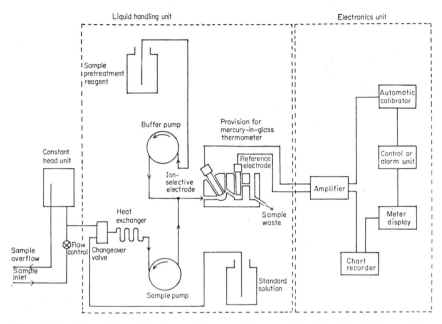

Fig. 9.10   The schematic flow diagram of an automatic industrial monitor
(E.I.L. Model 8000 monitor)

Industrial monitors are in common use for the analysis of fluoride, ammonia, nitrate and water hardness in various types of water; the floride monitor is also used in a control system with a dosing pump to regulate the fluoridation of water supplies. Other applications include the analysis of boiler feedwater with sodium and ammonium monitors and of industrial effluents with cyanide monitors.

REFERENCES

1. Orion Research Inc., *Newsletter* **2**, 42 (1970).
2. P. L. Bailey and E. Pungor, *Anal. Chim. Acta* **64**, 423 (1973).
3. G. Horvai, K. Tóth and E. Pungor, *Anal. Chim. Acta* **82**, 45 (1976). Budapest, 1975.
4. A. Hulanicki and W. Trojanowicz, *Anal. Chim. Acta* **68**, 155 (1974).
5. Orion Research Inc., *Newsletter* **2**, 5 (1970).
6. Orion Research Inc., *Newsletter* **1**, 25 (1969).
7. Orion Research Inc., *Newsletter* **1**, 9 (1969).

8. R. A. Durst, in *Ion-Selective Electrodes*, R. A. Durst, editor, Chapter 11. N.B.S. Spec. Publ. No. 314, Washington, D.C. (1969).

9. A. Hulanicki, R. Lewandowski and M. Maj, *Anal. Chim. Acta* **69**, 409 (1974).

10. R. F. Thomas and R. L. Booth, *Environ. Sci. Technol.* **7**, 523 (1973).

11. J. A. Krueger, *Anal. Chem.* **46**, 1338 (1974).

12. G. Gran, *Analyst* **77**, 661 (1952).

13. Orion Research Inc., *Newsletter* **2**, 49 (1970).

14. Orion Research Inc., *Newsletter* **3**, 1 (1971).

15. Orion Research Inc., *Newsletter* **3**, 17 (1971).

16. M. S. Frant, J. W. Ross and J. H. Riseman, *Anal. Chem.* **44**, 2227 (1972).

17. B. Fleet and A. Y. W. Ho, in *Ion-Selective Electrodes*, E. Pungor, editor, p. 1. Akadémiai Kiadó, Budapest (1973).

18. E. Bishop, in *Comprehensive Analytical Chemistry*, C. L. Wilson and D. W. Wilson, editors, Vol. IB. Elsevier, Amsterdam (1960).

19. D. E. Davis, in *Comprehensive Analytical Chemistry*, C. L. Wilson and D. W. Wilson, editors, Vol. IIA. Elsevier, Amsterdam (1964).

20. E. W. Baumann and R. M. Wallace, *Anal. Chem.* **41**, 2072 (1969).

21. J. W. Ross and M. S. Frant, *Anal. Chem.* **41**, 1900 (1969).

22. B. Fleet and A. Y. W. Ho, *Anal. Chem.* **46**, 9 (1974).

23. P. W. Carr, *Anal. Chem.* **43**, 425 (1971).

24. F. A. Schultz, *Anal. Chem.* **43**, 1523 (1971).

25. E. Eckfeldt, *Anal. Chem.* **47**, 2309 (1975).

26. E. Eckfeldt and W. E. Proctor, *Anal. Chem.* **47**, 2307 (1975).

27. P. Van den Winkel, J. Mertens and D. L. Massart, *Anal. Chem.* **46**, 1765 (1974).

28. A. L. Wilson, *The Chemical Analysis of Waters*, Society for Analytical Chemistry, London (1974).

29. L. E. Negus and T. S. Light, *Instrum. Technol.* 23 (December 1972).

30. P. L. Bailey and M. Riley, *Analyst* **100**, 145 (1975).

31. G. K. Buckee, *J. Inst. Brewing*, **80**, 291 (1974).

CHAPTER 10

MANUFACTURE OF ELECTRODES

The analyst may sometimes wish to make his own ion-selective electrodes, either for reasons of economy and convenience or because the particular electrode which he requires is not commercially available. The electrodes based on inorganic salts, organic ion exchangers and neutral carriers are all quite simple to make, provided that the appropriate active materials are available (see Tables 5.1, 6.2 and 6.3). The glass electrodes are even easier to make, as a satisfactory electrode can be manufactured from a bulb of the special cation-sensitive glass blown on to a stem of high-resistance glass. However, the major difficulty is that the special glass is not readily available and only the formulae of the less satisfactory glasses are freely accessible from the literature. Furthermore, if the glass is to be home-made, a furnace capable of continuous operation at 1500 °C is required. Since the prices of the sodium-, potassium- and ammonium-selective glass electrodes are usually much lower than those of the other electrodes, the incentive for the analyst to make them himself is correspondingly lower, except for special applications requiring the use of an electrode with an extraordinary geometry. Thus, in this chapter the description will be confined to the principles of electrode manufacture and examples of manufacturing methods for electrodes based on inorganic crystals, organic ion exchangers and neutral carriers.

A gas-sensing probe is relatively difficult to prepare from scratch, as a body must be made to hold the specially shaped pH electrode and reference electrode and, for the membrane probes, a membrane found which is appropriate to the particular determinand. Those readers who wish to attempt this should consult the references given in Chapter 7.

PRINCIPLES OF CONSTRUCTION

Ion-selective electrodes, in general, consist of an ion-selective membrane sealed into the end of an inert tube: the membrane is connected, directly or indirectly, on the inside of the tube, via an internal reference electrode, to a cable which leads to the plug which is inserted in the pH meter. Thus, the following aspects of the design need to be considered:

1. Type of cable and plug
2. Body material
3. Sealing methods—adhesives
4. Membrane manufacture
   (a) liquid membranes
   (b) pressed pellet and single-crystal membranes
   (c) silicone rubber membranes
   (d) PVC membranes
   (e) polythene membranes
5. Type of internal contact to the membrane
   (a) internal reference electrode + filling solution
   (b) solid internal contact

Each of these aspects will be discussed in turn.

1. The impedances of many of the membranes lie between 1 and 100 MΩ and, hence, it is necessary to use a voltmeter with a high-impedance input (i.e. a pH meter) to measure the potentials across them. The current drawn by these meters from the electrode is typically in the range 10^{-13}–10^{-15} A, and, hence, the lead is extremely liable to pick up electrical noise from surrounding a.c. mains equipment unless a coaxial antimicrophonic cable is used.

The most satisfactory low-noise cables are those which have a copper braid, as screen, covering an antimicrophonic conductive layer coated on to the insulation of the central conductor (see Fig. 10.1). The layer, which usually

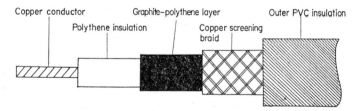

Fig. 10.1 Low-noise cable for electrodes

consists of graphite-impregnated polythene, minimizes the amount of static charge generated as a result of rubbing of the braid of the screen against the central insulation when the cable is bent. When the cable is stripped for fitting the plug, it is essential to clean this antimicrophonic layer off all that part of the cable core which is exposed by removal of the braid: the white central insulator should be cleaned with a solvent such as methylated spirits to remove the final traces of the conductor. If care is not taken at this stage, the core of the cable will not be adequately insulated from the screen (often earthed) after the plug is fitted.

The choice of plug will usually be dictated by the make of pH meter with which the electrode is to be used. There is a lamentable lack of standardization of the input sockets of pH meters, with at least six entirely different types in common use in Europe at the moment. In the U.K. the coaxial standard socket is by far the most common on indigenous laboratory meters; while B.N.C. and P.E.T. sockets, which are more protected, are preferred on industrial meters. Whenever possible after fitting a plug, it is wise to check the insulation resistance between the screen and core, if a suitable meter is available; the resistance should be at least 10^{12} Ω.

2. The body of the electrode must be made from a chemically inert material, and preferably from a material which is readily available. Many electrodes have been made with glass stems and this is certainly convenient, since glass tubing of an appropriate diameter is often to hand in an analyst's laboratory and many adhesives will stick to glass: lead glass is generally the most satisfactory. The electrode is less destructible if made with a plastics stem. Tubes of appropriately inert materials such as polypropylene, high-density polythene, PTFE, PCTFE or TPX (a methyl pentene polymer) cannot be used in conjunction with adhesives, as adhesives will not stick to them: the membrane has to be held in the stem by a compression seal as described in the next section. If a slightly less inert stem can be tolerated, PVC is a good choice, since many adhesives will stick to it.

3. It is at once one of the most difficult and the most important parts of electrode design to ensure good sealing. The two areas which are most vulnerable to sealing problems are the junction between the membrane and the body and, in electrodes with an internal filling solution, the junction of the internal reference electrode with the cable. In the first case leakage round the membrane may cause the membrane to be effectively shorted out by electrical contact between the sample and the internal filling solution or internal membrane contact. In the second case, if the internal solution or moisture evaporating from it reaches the end of the coaxial cable, this will cause tracking between the screen and the core of the cable and, hence, erratic response: moreover, if the internal solution continually contacts either the copper core or the joint between this and the internal reference electrode (usually a soldered joint, but it may also be, for example, a crimped joint), the internal reference potential will be shifted and drift as the solution reacts with the copper, tin, lead, etc.

In home-made electrodes the seals may be made most simply by means of a chemically resistant adhesive such as epoxy cement (e.g. one of the 'Araldite' family produced by CIBA-Geigy Ltd.) or a silicone rubber sealant (e.g. an R.T.V. adhesive produced by Dow Corning, I.C.I. or the General Electric Co., N.Y.). The epoxy cement gives a strong and rigid seal if carefully applied in a uniform film and if the two surfaces to be joined are clean: it is particularly suitable for fixing the membrane into the end of the electrode body. Some of the epoxy cements available are unpleasant to handle and can cause inflammation of the skin; hence, it is advisable to wear gloves and to work in a well-ventilated area when weighing out, mixing and using them. The silicone rubber sealant may be used for encapsulating the cable in the electrode body or, at least, the end of the cable from above where the braid is exposed down to below the joint with the internal reference electrode. It is advisable to avoid using those types of silicone rubber which liberate acetic acid on curing, unless it is certain that the acid will not harm the membrane, cable or internal reference electrode. Both adhesives take several hours to cure fully and should not be wetted until cured.

As an alternative to the use of an adhesive to hold the membrane in place, the end of the electrode stem may be threaded to take an end-cap and the membrane held inside this. Exactly such an arrangement is used to hold the membrane in a gas-sensing membrane probe, as shown in Fig. 7.1. The seals in this case are made by the 'O'-ring and the internal sealing washer; a suitable material for these is fluorinated synthetic rubber or another inert elastic polymer. If compression seals are used in this way, the body may be machined from one of the inert plastics mentioned in the previous section.

4. The choice must first be made of the best active material to be used in the membrane; reference to the tables in Chapters 5 and 6 should give some guidance as to the more successful materials which have been tried. Then the form of the membrane must be selected; in the following paragraphs some guidance is given on methods of manufacturing the different membranes.

(a) The ion-selective electrodes which are based on organic ion-exchangers or neutral carriers may easily be constructed with liquid membranes. The two simplest constructions are that of the Corning-type electrodes and the electrode of Hulanicki, Lewandowski and Maj:[1] the designs of these are shown in Figs 6.1 and 10.2, respectively.

In the Corning electrode a ceramic plug is sealed into the bottom of the electrode body. This ceramic plug may take the form of a fine glass frit or be of the type used for forming liquid junctions in reference electrodes: in either case the plug should be made hydrophobic before use by treatment with one of the commercial silicone coating compounds. Alternatively, if this junction is to be held in place by an end-cap, instead of being sealed into the body, a hydrophobic cellulose acetate membrane may be used (e.g. a 'Millipore' filter), as shown in Fig. 6.1. The solution of the active material in the appropriate solvent (see Tables 6.2 and 6.3) is poured into the body and seeps through the ceramic to form a sharp boundary with the sample. The layer of the solution nearest to the edge of the ceramic may be renewed, after the electrode has been in use for some time or after short-term storage, by dabbing the end with cotton wool; this will absorb the surface layer of solution and draw more fresh solution to the boundary to replace it. The internal reference electrode may be made by inserting into the solution a tube containing a silver/silver chloride electrode of the second kind and an agar gel of a solution containing chloride ions.

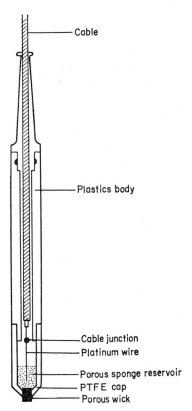

Fig. 10.2   An ion-selective electrode with a liquid membrane[1]

The electrode of Hulanicki *et al.*,[1] shown in Fig. 10.2, is different in three major respects. First, the ceramic plug is replaced by a porous wick of a 'natural or synthetic porous polymer'; presumably any wick which is resistant to the solution of active material and to the samples would suffice. Second, the internal contact with the active material is made not by means of the usual reference electrode but directly by dipping a platinum wire into it. In the example in the paper[1] a nitrate electrode was constructed, but it is not clear, and is even doubtful, whether this arrangement with the platinum wire would work satisfactorily with other active materials, since there is no apparent reversible redox reaction occurring at the electrode surface to stabilize the internal reference potential. However, if the platinum wire were replaced by a conventional internal reference system, similar to that used in the Corning electrode, the design could be widely used. Third, the internal reservoir of the solution of active material is soaked into a sponge of polyurethane foam in contact with the top of the wick; the solution is thus immobilized within the electrode so that the electrode may be laid down or used in the horizontal position without detriment.

(b) Solid membranes consisting entirely of the active material are commonly used in the electrodes based on inorganic salts. They are prepared in two principal ways: by growing a single crystal or by making a pellet of the active material. Which way is chosen depends on the material in question.

The fluoride electrode is the electrode most commonly made with a single crystal as membrane, in this case a crystal of lanthanum fluoride doped with europium fluoride (although doped crystals are normally used, the doping is probably not essential). It is seldom worth while going to the trouble of growing these crystals, however, since they are readily available commercially at reasonable cost. All that remains to be done therefore is to fix the crystal into the body and to arrange the internal reference system.

Pellets of inorganic salts for use as membranes may be prepared in pellet presses designed for spectroscopic work: a 15 ton press with an evacuable die which can be heated to 200 °C is suitable for all purposes (such presses are obtainable in the U.K. from, among others, Spectroscopic Accessories Ltd). The conditions of pressing vary according to the material being pressed but, in general, pressing under vacuum results in denser, less porous membranes than those formed at atmospheric pressure.[2] The heating facility is required in the manufacture of only a few pellets such as the mercurous chloride/mercuric sulfide pellet.[3]

The method of preparation of the material for pressing has a strong influence on the properties of the resultant electrode. Usually the pellets with the best mechanical and chemical properties are prepared from mixtures of compounds; it is important to ensure that powder to be pressed is an intimate mixture of its components and is finely divided.[2] The intimate mixture of the components is best achieved by coprecipitation. Thus, a method for preparation of a silver chloride/silver sulfide pellet has been described[2] as follows. A solution is made up, containing approximately equal molar concentrations of sodium chloride and sodium sulfide, and added to a solution containing a double excess of silver nitrate. A coprecipitate of silver chloride and silver sulfide is formed, which is filtered off and thoroughly washed to remove all traces of nitrate ion by 20 or more rinses with distilled water. The precipitate is then dried at 110 °C and is ready for compression. A similar procedure has been used in the preparation of the active material for the other halide electrodes[2] and also the copper, lead and cadmium electrodes.[4]

(c) Silicone rubber has been used as the inert matrix for a wide range of active materials. The original electrodes with silicone rubber membranes, produced by Pungor and his co-workers, were designed for measurement of the halides, pseudohalides and silver ions, but since then they and others have extended the range of determinands which may be measured by electrodes with silicone rubber membranes to include potassium ions, copper ions, surfactants and others.

The preparation of these membranes is quite straightforward and has been described by both Pick et al.[5] and Fogg, Pathan and Burns,[6] although the potassium electrode of Pick *et al.* is unusual in the small quantity of active material used. The active material should be a fine powder (particle size 1–20 μm); the correct size may be assured by using an appropriate precipitation technique to prepare the material and/or by grinding the powder. In the preparation of the silver halides used in the halide-selective electrodes, a halide solution is titrated with silver nitrate solution in the presence of *p*-ethoxychrysoidine to prevent secondary nucleation; a small excess of the silver solution is added, after the end-point is reached, to ensure an excess of silver ions on the surface of the precipitate,[7] as in the preparation of the silver halides to be pressed into pellets.[2] These positively charged precipitates were found by Pungor[8] to be most satisfactory from the point of view of both membrane conductivity and performance, although it was shown that the exchange rate of iodide membrane and the sample is increased if the precipitate is negatively charged. To prepare the active material for a copper-selective electrode, Pick, Tóth and Pungor[9] precipitated copper sulfide from a homogeneous solution of copper ions and thiosulfate ions at pH 0; the average particle size of the precipitate was found to be 1.5 μm.

Once the active material has been prepared, it is thoroughly mixed with the polysiloxane (unpolymerized silicone rubber); usually the proportion of active material to polysiloxane is made as large as possible, so that the proportion of active material in the final membrane may be as high as 75% w/w. A small quantity of the cross-linking agent + catalyst reagent is then mixed in (the quantity according to the manufacturer's instructions) and the rubber is spread on a clean glass plate. A second plate is placed on top, separated at the four corners by spacers about 1 mm thick (coins are suitable) and kept down by a weight. The rubber is then allowed 24 h to cure. Membranes may be cut from the resultant sheet with a cork borer and stuck on to the end of glass tubes with a silicone rubber sealant. The membranes require conditioning by leaving them to soak for about two hours in a solution of the determinand (about 10^{-2} M is suitable) before use.

In the exceptional case of the silicone rubber electrode containing valinomycin,[5] the proportion of valinomycin in the membrane was only 5% w/w.

(d) PVC has primarily been used as an inert matrix for the organic ion exchangers and also for the neutral carriers. A full description of the preparation of calcium- and nitrate-selective electrodes with PVC membranes has been published by Craggs, Moody and Thomas.[10] The membrane is prepared by dissolving PVC (0.17 g) in tetrahydrofuran (6 ml) and then thoroughly stirring in a solution of a commercial ion exchanger (0.40 g). The solution is poured into a glass ring (30–35 mm diameter) sitting flush on a glass plate: a pad of filter papers and a heavy weight are then placed on top of the ring and the assembly (Fig. 10.3) is left for two days to allow the tetrahydrofuran to evaporate. The glass ring is then removed and the sheet of membrane material prised away. Alternatively, the solution may be poured into a Petri dish for the evaporation to occur.[11] The individual membranes may be cut from the sheet with a cork borer and stuck on to a flat-ended PVC tube with a solution of PVC in tetrahydrofuran. The membrane is usually approximately 0.5–1.5 mm thick.

Satisfactory potassium-selective electrodes[12] have been prepared from a solution of valinomycin (5 mg) in dipentyl phthalate (100 mg) in a matrix of PVC (45 mg). The membrane is prepared as described for the calcium and nitrate electrodes.

(e) Polythene is a satisfactory alternative to silicone rubber or PVC as the inert matrix. A review of the preparation and development of ion-selective electrodes with polythene membranes has been presented by Liberti.[13] As an example, the preparation of an iodide electrode is described. A precipitate of silver iodide is formed by addition of an excess of 0.5 M silver nitrate solution to a 10^{-2} M potassium iodide solution. The precipitate is filtered off, washed, dried and ground to a fine powder together with a fine powder of polythene

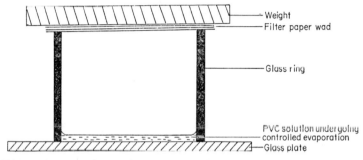

Fig. 10.3   Experimental arrangement for casting PVC membranes[10]

(approximately two parts by weight of precipitate to one of polythene). The mixture is placed in a press and heated to 100–130 °C under a pressure of 100–300 atm (1–3×10^7 Pa), and then left to cool. Enough of the mixture should be used to produce a membrane 1–2 mm thick. The membrane may be fixed on to the end of a rigid polythene tube by heat treatment, which gives a strong and durable seal. Electrodes have been prepared in this way which are selective to the halides, pseudohalides, copper, lead, cadmium and silver ions.[13]

5. Once the membrane is in place, the next stage is to connect the back of the membrane to the cable in such a way that any potentials developed between the various phases are stable. As already described at the beginning of Chapter 5, there are two ways in which this may be achieved. Either an internal filling solution and internal reference electrode is used or the internal wire is fixed directly to the back of the membrane: the second alternative is only available with a limited number of membranes.

(a) An internal filling solution should contain ions to which both the membrane and the internal reference electrode give a thermodynamically reversible response. In practice, since the most convenient internal reference electrode is a silver/silver chloride wire, this means that the solution contains determinand ions and chloride ions, usually both at a level of about 10^{-1}–10^{-2} M. Unless an excess of solid silver chloride is added to the solution, it is inadvisable to increase the chloride concentration above this level, as otherwise the silver chloride coating on the reference electrode is rapidly stripped off if the temperature increases. The concentration of determinand ions is not critical for the membranes based on inorganic salts; however, for the other membranes, if the concentration of determinand in the filling solution is very different from that in the samples, drift occurs.[14] Thus, the internal filling solution of an

TABLE 10.1
Some methods of preparation of ion-selective electrodes

Electrode	Membrane +active material	Internal filling solution	Internal reference electrode	Comments	Ref.
F^-	single crystal of LaF_3 doped with EuF_2	1 M KF+sat. KCl +sat. AgCl	Ag/AgCl	The doping of the crystal with EuF_2 is not essential. No other type of membrane is as satisfactory.	15
NO_3^-	liquid membrane+ NO_3^- active material (a) in Table 6.2	10^{-1} M NaCl+ 10^{-2} M NaNO_3+ sat. AgCl	Ag/AgCl		
Ca^{2+}	PVC membrane+Ca^{2+} active material (c) (DOPP) in Table 6.2	10^{-2} M $CaCl_2$+ sat. AgCl	Ag/AgCl	Alternative active materials are described in Refs 17 and 18; preparative details for active material in Refs 16 and 17.	16
K^+	liquid membrane 0.009 M valinomycin in diphenyl ether/'Millipore' filter or	10^{-2} M KCl+ sat. AgCl	Ag/AgCl	Valinomycin is commercially available.	19
	silicone rubber membrane incorporating valinomycin	10^{-2} M KCl+ sat. AgCl	Ag/AgCl	Easier to make.	5
Cl^-	$AgCl/Ag_2S$ pellet	10^{-1} M $AgNO_3$	Ag/AgCl	Precipitate active material with excess Ag^+.	2
Cu^{2+}	Cu_xS/Ag_2S pellet	10^{-3} M $Cu(NO_3)_2$ +10^{-2} M NaCl +sat. AgCl or solid Ag contact	Ag/AgCl		4

Orion fluoride electrode[15] consists of 1 M potassium fluoride saturated with potassium chloride and silver chloride, whereas the internal filling solution of an Orion calcium electrode consists of 0.14 M sodium chloride + 0.01 M calcium chloride.[16]

In the electrodes with solid membranes the filling solution may be either gelled or absorbed into a sponge on the back of the membrane. This helps to keep the membrane moist all the time and also stops the solution running out of the top of the electrode or round the junction between the internal reference electrode and the cable when the electrode is laid down.

(b) Unfortunately, only those electrodes having silver salts in the membrane may be successfully prepared in a completely solid state version giving stable potentials. A simple method of constructing these electrodes, is to fix a silver wire on to the back of the membrane using a silver-loaded epoxy cement, and then solder the silver wire to the electrode cable. The membrane is sealed into a suitable electrode body to give an electrode of the type shown in Fig. 5.1. Because there is no solution inside the electrode, the internal sealing problems disappear completely, leaving only the seal between the membrane and the body to be made. Hence, this construction is to be preferred whenever it is applicable, which, as shown in Table 5.1, is in electrodes selective to the halides, pseudo-halides, copper, lead, cadmium and silver ions.

The theory associated with the development of stable potentials across these solid contacts was described in Chapter 5.

From the foregoing descriptions it will be clear that electrodes sensing some ions, notably the halides and heavy metal ions, may be made in several ways. Often there is very little to choose between the performances of electrodes made in these different ways when they are based upon the same active material, and, hence, the most convenient method of preparation may be used. A summary of some of the many possible methods of preparation of the more popular electrodes is given in Table 10.1: many of the successful designs which have been commercially exploited are patented.

REFERENCES

1. A. Hulanicki, R. Lewandowski and M. Maj, *Anal. Chim. Acta*, **69**, 409 (1974).
2. J. W. Ross, J. H. Riseman and M. S. Frant, U.S. Patent No. 3,563,874 (16th February 1971).
3. J. F. Lechner and I. Sekerka, *J. Electroanal. Chem.* **57**, 317 (1974).
4. M. S. Frant and J. W. Ross, U.S. Patent No. 3,591,464 (6th July 1971).
5. J. Pick, K. Tóth, E. Pungor, M. Vasák and W. Simon, *Anal. Chim. Acta* **64**, 477 (1973).
6. A. G. Fogg, A. S. Pathan and D. T. Burns, *Anal. Chim. Acta*, **69**, 238 (1974).
7. J. Havas, E. Papp and E. Pungor, *Acta Chim. Acad. Sci. Hung.* **58**, 9 (1968).
8. E. Pungor, *Anal. Chem.* **39** (13), 28A (1967).
9. J. Pick, K. Tóth and E. Pungor, *Anal. Chim. Acta* **61**, 169 (1972).
10. A. Craggs, G. J. Moody and J. D. R. Thomas, *J. Chem. Educ.* **51**, 541 (1974).

11. M. Mascini and F. Pallozzi, *Anal. Chim. Acta* **73**, 375 (1974).
12. W. E. Morf, E. Lindner and W. Simon, *Anal. Chem.* **47**, 1596 (1975).
13. A. Liberti, in *Ion-Selective Electrodes*, E. Pungor, editor, p. 37. Akadémiai Kiadó, Budapest (1973).
14. J. N. Butler, in *Ion-Selective Electrodes*, R. A. Durst, editor, Chapter 5. N.B.S. Spec. Publ. No. 314, Washington, D.C. (1969).
15. M. S. Frant, U.S. Patent No. 3,431,182 (4th March 1969).
16. J. Růžička, E. H. Hansen and J. C. Tjell, *Anal. Chim. Acta* **67**, 155 (1973).
17. R. W. Cattrall, D. M. Drew and I. C. Hamilton, *Anal. Chim. Acta* **76**, 269 (1975).
18. R. W. Cattrall and D. M. Drew, *Anal. Chim. Acta* **77**, 9 (1975).
19. L. A. R. Pioda, V. Stankova and W. Simon, *Anal. Lett.* **2**, 665 (1969).

SUBJECT INDEX

A

Acetylcholine electrode, 178
Activity coefficients, 35–43, 200
  determination of, 72
Activity, ionic, 2, 35–40
Activity standards, 38–40
Air-borne particulates, analysis of, 74, 141, 164
Air-gap electrode, 4, 147 167
Albuminoid nitrogen, determination of, 161
Aluminium, determination of, 113
Ammonia/ammonium, determination of, 144, 161, 208
  in air-borne particulates, 74, 164
  in blood, 164
  in boiler feed-water, 74, 144, 162
  in Kjeldahl digests, 154, 159, 163, 201, 209
  in pharmaceuticals, 164
  in various waters, 159, 161–162, 190
  reference electrode bridge solution, 29
Ammonia probe, 9, 74, 132, 141, 144–145, 147, 149–165, 170, 208–209
Ammonium electrode (based on neutral carriers), 74, 117, 122, 127, 128, 132, 144–145, 162
Ammonium, potassium and general cation electrode (glass), 60–62, 73–74, 132, 144, 162, 170
Analate addition method, 190
Asymmetry potential, 61–62

B

Bacon, analysis of, 72
Beer, analysis of, 138, 144, 164
Biological fluids, analysis of, 70, 73, 136, 141
Blood, analysis of, 2, 40, 136, 147, 164, 165
Boiler feed-water, analysis of, 55, 70, 74, 106, 107, 144, 162, 165, 203, 204
Boron, determination in tap water, 143
Bridge solutions, 11–12, 23–30
  equitransferent solutions, 27–29
  in Ag/AgCl reference electrodes, 18–20
  KCl solutions, 14, 15, 27–28
Bromide, determination of, 108–109
Bromide electrode, 76, 79, 104, 108–109
Buffers for electrode calibration, 41–43

C

Cable, 215
Cadmium electrode, 76, 79, 82–83, 87–89, 97–99, 113
  applications, 113
Calcium, determination of, 136–139
  in beer, 138
  in blood, 2, 40, 136
  in milk, 137
  in soda-lime, 138
  in waters, 137, 187
Calcium electrode, 8, 117, 120–122, 124–130, 132–134, 136–139